W9-CPY-330

The Canadian System of Soil Classification

Soil Classification Working Group

NRC Research Press
Ottawa 1998

Research Branch
Agriculture and Agri-Food Canada
Publication 1646
Third Edition 1998

Printed in Canada on acid-free paper

ISBN-13: 978-0-660-17404-4
ISBN: 0-660-17404-9
NRC No. 41647

Canadian Cataloguing in Publication Data

Canada. Soil Classification Working Group

The Canadian system of soil classification

3rd ed.
"Research Branch, Agriculture and Agri-Food Canada Publication"
Issued by the National Research Council of Canada.
Includes bibliographical references and an index.
"The changes incorporated in this current publication are based on the work of the Soil Classification Working Group formerly of the Expert Committee on Soil Survey..." Pref.
ISBN 0-660-17404-9   ISBN-13 978-0-660-17404-4

1. Soils — Canada — Classification. 2. Soils — Classification.
I. Canada. Agriculture Canada. Research Branch.
II. National Research Council of Canada. III. Title.

S599.1S54   1998   631.44'0971   C98-980032-6

First printed 1998
Second printing 2003
Third printing 2005
Fourth printing 2007

NRC Monograph Publishing Program

Inquiries: Monograph Publishing Program, NRC Research Press, National Research Council of Canada, Ottawa, Ontario K1A 0R6, Canada.

First published in 1974 under title *A System of Soil Classification for Canada*, Agric. Can. Publ. 1455.

Superseded in 1978 by *The Canadian System of Soil Classification*, Agric. Can. Publ. 1646.

Second edition 1987, *The Canadian System of Soil Classification*, Agric. Can. Publ. 1646.

Third edition 1998, correct citation as follows:

Soil Classification Working Group. 1998. *The Canadian System of Soil Classification.* Agric. and Agri-Food Can. Publ. 1646 (Revised). 187 pp.

Contents

Figures

Tables

Preface

This revised publication replaces *The Canadian System of Soil Classification* (second edition) published in 1987. The changes incorporated in this current publication are based on the work of the Soil Classification Working Group formerly of the Expert Committee on Soil Survey, and continued by the Land Resource Division of the former Centre for Land and Biological Resources Research, Ottawa, Ont.

Major revisions have been carried out to the Cryosolic order; a tenth order, Vertisolic, has been added; subgroups intergrading to the Vertisolic order have been added in the Chernozemic, Gleysolic, Luvisolic, and Solonetzic orders; and a complete *Key to Soil Classification* has been added in Chapter 3. The main reason for modifying the Cryosolic order stems from the work of Charles Tarnocai, Ottawa, and Scott Smith, Whitehorse, who assisted the International Gelisol Working Group in establishing the Gelisol order for *Soil Taxonomy*. The most significant changes include adding several new subgroups in both the Turbic Cryosol and Static Cryosol great groups and revising the description of all subgroups to make them as uniform as possible and to clearly identify those properties diagnostic of the particular subgroup.

The main initiative for establishing the Vertisolic order resulted from the decision by the United States Department of Agriculture to establish a new suborder *Cryert* in the Vertisol order of *Soil Taxonomy* that allows clay soils with Vertic properties that occur in cold climates to be classified as Vertisols. In light of this decision, a review of the classification of clay soils in Canada by the Soil Classification Working Group led to the recommendation to establish the Vertisolic order.

In addition to the revisions described above, a number of errors and discrepancies in the 1987 edition of *The Canadian System of Soil Classification* have been corrected. As well, the wording has been changed in many instances to remove ambiguity, to standardize terminology, and to make the intent more easily understood.

Most of the work relating to these changes was carried out by members of the Soil Classification Working Group: H.A. Luttmerding, British Columbia; J.A. Brierley, Alberta; A.R. Mermut and H.B. Stonehouse, Saskatchewan; W. Michalyna, Manitoba; K.B. MacDonald, Guelph, Ont.; J.A. Shields and C. Tarnocai, Ottawa, Ont.; L. Lamontagne, Quebec; K.T. Webb, Nova Scotia; and by other pedologists across Canada: C.A. Smith, Yukon; R.G. Eilers and Hugo Veldhuis, Manitoba; H.W. Rees, New Brunswick; D.A. Holmstrom, Prince Edward Island; and E.F. Woodrow, Newfoundland.

The members of the Working Group wish to express their appreciation to all other persons, both active and retired, who cooperated in this work. They would also like to thank all those who worked behind the scenes to help arrange and conduct Soil Correlation Tours that aided in establishing the Vertisolic order and revising the Cryosolic order. Also, special thanks are extended to C. Wang, Ottawa, for his work on the *Key to Soil Classification*.

Questions and suggestions on classification are welcomed and should be sent to the Program Manager, Land Resource Evaluation, Eastern Cereal and Oilseed Research Centre, Ottawa, Ont., K1A 0C6.

H.B. Stonehouse
Coordinator
Soil Classification Working Group

Preface to the first edition (1978)

This publication replaces *The System of Soil Classification for Canada* published by Canada Department of Agriculture as Publication 1455, Revised 1974. Some of the changes in the system were agreed upon at the 1973 and 1976 meetings of the Canada Soil Survey Committee (CSSC) and subsequent decisions have been made by the Subcommittee on Soil Classification.

The main changes are as follows:

1. The inclusion of a Cryosolic order for soils having permafrost close to the surface.
2. The elimination of subgroup modifiers and therefore a reduction in the number of subgroup combinations.
3. The deletion of soil type as a category in the system.
4. The increased uniformity of presentation of the soil orders.
5. The amplification of the introductory material to give more of the background and rationale of soil classification in Canada.

This publication presents the current state of soil taxonomy in Canada. The Canadian system was influenced by history, regional biases, new information on soils in Canada and elsewhere, and international concepts of soil. It represents the views of most Canadian pedologists, but not all the details are satisfactory to any one pedologist. This is a stage in the evolution of an improved system that will result from further knowledge of soils and an improved arrangement of the information.

The history and rationale of soil classification in Canada are outlined briefly to point out the changes in concepts and the current point of view on soil taxonomy. This is followed by chapters that define soil, soil horizons, and other basic terms, and explain how to key out the classification of a soil. A chapter is devoted to each of the nine soil orders and the great groups and subgroups within each order. The orders are arranged alphabetically, but great groups and subgroups are arranged as they were in previous versions of the system. Chapters on the family and series categories and on soil phases follow. The recently developed landform classification system for soil surveys that was approved by CSSC in 1976 is included as a separate chapter. For additional information see: Canada Department of Agriculture (1976), Clayton et al. (1977), Working Group on Soil Survey Data (1975), and soil survey reports available from provincial soil survey units.

This revision was prepared by the Subcommittee on Soil Classification of the Canada Soil Survey Committee: T.M. Lord, British Columbia; W.W. Pettapiece, Alberta; R.J. St. Arnaud, Saskatchewan; R.E. Smith, Manitoba; C.J. Acton, Ontario; R. Baril, Quebec; G.J. Beke and C. Wang, Atlantic Provinces; and J.H. Day, J.A. McKeague, J.L. Nowland, and J. A. Shields, all of Ottawa.

The Subcommittee wishes to thank all those who shared in the preparation of this publication. Many pedologists critically reviewed drafts of sections and made useful suggestions. Janet Lyons typed several drafts of the manuscript. B. Baker, Graphics Section, Research Program Service, prepared the drawings.

Readers are invited to send their questions and suggestions to the chairman of the Canada Soil Survey Committee, Soil Research Institute, Agriculture Canada, Ottawa, Ont.

J. A. McKeague
Chairman
Subcommittee on Soil Classification

Preface to the second edition (1987)

This revised publication replaces *The Canadian System of Soil Classification* published in 1978. The changes incorporated in the present publication are based on the work of the Soil Classification Working Group of the Expert Committee on Soil Survey.

Major revisions have been carried out on both the Gleysolic and Organic orders to incorporate the new information obtained on these soils and to overcome difficulties in their classification under the earlier system. Soil scientists from all over Canada requested that the classification of Gleysols be reviewed. This review resulted in the revision of the basic rationale commonly used in the past by which all poorly drained soils were identified as Gleysols. The revised classification of the Gleysolic soils presented here is based on soil properties that indicate reduction during the genesis of the soil and not on the current soil water regime. Although many Gleysolic soils are poorly or very poorly drained, some are imperfectly drained and others, in which drainage has been modified, are now well drained. Because of these changes, the section on the Gleysolic order now has a slightly longer introduction.

The main initiative for reviewing the Organic order, especially the Folisol great group, came from British Columbia, where most of these soils are found. The importance of organic soil layers in forest ecosystems has been emphasized by foresters and forest soil specialists working in that province. This view is reflected in the revised Folisol great group. With this change the Folisol great group and the other great groups of the Organic order now have a common basis for classification.

The revision of the Podzolic order was minor and was carried out to make the definitions more specific. The new definition for contrasting horizons and layers was made more detailed in an attempt to avoid the misinterpretations that occurred in the past.

In addition to the revisions described, a number of errors in the 1978 edition of *The Canadian System of Soil Classification* have been corrected.

Most of the work relating to these changes was carried out by members of the Soil Classification Working Group: T.M. Lord and R. Trowbridge, British Columbia; L. Turchenek, Alberta; R.J. St. Arnaud, Saskatchewan; W. Michalyna and G.F. Mills, Manitoba; L.J. Evans, Ontario; J-M. Cossette, Quebec; K.T. Webb, Nova Scotia; C.A.S. Smith, Yukon Territory; and J.A. McKeague, C.A. Fox, J.A. Shields, and C. Tarnocai, Ottawa, Ont.

The members of the Working Group wish to express their appreciation to all those who cooperated in this work, with special thanks to H.A. Luttmerding for his work on Gleysols and Folisols and to C. Wang for his work on the Gleysolic and Podzolic soil classification.

Questions and suggestions on soil classification are welcomed and should be sent to the chairperson of the Soil Classification Working Group, Expert Committee on Soil Survey, Land Resource Research Centre, Agriculture Canada, Ottawa, Ont., KlA 0C6.

C. Tarnocai
Chairperson
Soil Classification Working Group

Chapter 1

Introduction

History of Soil Classification in Canada

The early years, 1914–1940

Classifying soils in Canada began with the first soil survey in Ontario in 1914. When A. J. Galbraith set out to map the soils of southern Ontario, concepts of soil and methods of soil classification were rudimentary in North America. G.N. Coffey, formerly of the U.S. Bureau of Soils, advised Galbraith during the early stages of the survey. The system of classification used was that of the U.S. Bureau of Soils, which was based largely upon geological material and texture (Ruhnke 1926). Nine "soil series" were mapped in all Ontario south of Kingston by 1920. The broad scope of these "series," which were somewhat analogous to geological formations at that time, has narrowed progressively to the present.

Changes in the system of classifying the soils of Canada have resulted from the combined effects of international developments in concepts of soils and increasing knowledge of Canadian soils. From the time of the first surveys Canadian pedologists were influenced by the concept of soil as a natural body integrating the accumulative effects of climate and vegetation acting on surficial materials. This concept was introduced by Dokuchaev about 1870, developed by other Russian soil scientists, and proclaimed to western Europe by Glinka in 1914 in a book published in German. Marbut's translation of this book made the Russian concept of soil as a natural body easily available to the English-speaking world (Glinka 1927). The concept is of paramount importance in soil science because it makes it possible to classify soils on the basis of properties of the soils themselves rather than on the basis of geology, climate, or other factors. Classification systems based on the inherent properties of the objects classified are called natural or taxonomic systems.

Recognizing the relationships between soil features and factors of climate and vegetation was not limited to Russian scientists. In the

United States, Hilgard had noted this association in a book published in 1860 (Jenny 1961), and Coffey had recognized soils as natural bodies by 1912 (Kellogg 1941). However, the Russians can be credited with developing the concept of soils as natural bodies with horizons that reflect the influences of soil-forming factors, particularly climate and vegetation.

Although classification was on the basis of texture in the earliest soil surveys undertaken in the Prairie Provinces during the 1920s, an increased awareness of soil zones and of the soil profile is evident in soil survey reports published during that decade. Preliminary soil zone maps of Alberta and Saskatchewan were presented by Wyatt and Joel at the first International Congress of Soil Science in 1927. They showed the broad belts of brown, black, and gray soils. The Congress and the associated field tours brought Canadian pedologists into close contact with international concepts of soil and systems of soil classification.

Developments in soil classification occurred independently in each province because surveys were carried out by university departments of soils or chemistry. For example, a numbering system indicating the soil zone, nature of parent material, mode of deposition, profile features, and texture was developed in Alberta. J.H. Ellis in Manitoba recognized the impossibility of developing a scientific soil taxonomy based on the limited knowledge of Canadian soils in the 1920s. Influenced by concepts of C.C. Nikiforoff in Minnesota, Ellis developed a field system of soil classification that was useful in soil mapping and endures in various revised forms to this day. The system identified "associations" of soils formed on similar parent materials and "associates" that differed according to topographic position within the association (Ellis 1932, 1971).

During the 1930s soil surveys proceeded in Ontario and the Prairie Provinces and were started in British Columbia in 1931, Quebec and Nova Scotia in 1934, and New Brunswick in 1938. Soil surveys began in Prince Edward Island in 1943, the Northwest Territories and the Yukon in 1944, and Newfoundland in 1949. The few soil surveyors employed permanently by federal and provincial departments of agriculture in the 1930s worked cooperatively with personnel of university soils departments. By 1936 about 15 000 000 ha (1.7% of the land area of Canada) had been surveyed, mainly in Alberta, Saskatchewan, and Ontario. Soil classification was limited by the fragmentary knowledge of the soils of Canada.

Canadian pedologists were influenced in the 1930s by Marbut's developing ideas on soil classification, Ellis's system of field classification (Ellis 1932), and the system of classification used in the U.S. Department of Agriculture (USDA) described by Baldwin et al. (1938). The latter system divided soils at the highest category among three orders:

1. Zonal soils, which are those with well-developed characteristics that reflect the influence of active factors of soil genesis such as climate, organisms, and particularly, vegetation (e.g., Podzol).

2. Intrazonal soils, which are soils having more or less well-defined characteristics that reflect the dominant influence of some local factor of relief or parent material over the normal effects of climate and vegetation (e.g., Humic Gleysol).

3. Azonal soils, which are soils without well-developed characteristics due either to their youth or to some condition of relief or parent material (e.g., alluvial soils).

Zonal soils were divided at the suborder level on the basis of climatic factors, and suborders were subdivided into great groups that were more or less similar to the great groups of today.

Canadian experience showed that the concept of zonal soils was useful in the western plains, but was less applicable in eastern Canada. In eastern Canada parent material and relief factors had a dominant influence on soil properties and development in many areas. However, the 1938 USDA system was used in Canada, and it influenced the subsequent development of the Canadian system.

From 1940 to 1996

The formation of the National Soil Survey Committee of Canada (NSSC) was a milestone in developing soil classification and pedology, generally, in Canada. The initial organizational meeting was held in Winnipeg in 1940

by the Soils Section of the Canadian Society of Technical Agriculturists (Ellis 1971). Subcommittees were established to prepare reports on six major topics, including soil classification. At the suggestion of E.S. Archibald, Director of the Experimental Farms Service, the NSSC became a committee of the National Advisory Committee on Agricultural Services. The first executive committee of the NSSC consisted of: A. Leahey, chairman; P.C. Stobbe, secretary; F.A. Wyatt, western representative; and G.N. Ruhnke, eastern representative. Terms of reference for the NSSC were developed by A. Leahey, G.N. Ruhnke, and C.L. Wrenshall. They were modified and restated in 1970 by the Canada Soil Survey Committee (CSSC) as follows:

To act as a coordinating body among the soil survey organizations in Canada supported by the Canada Department of Agriculture, provincial departments of agriculture, research councils, and departments of soil science at universities. Its stated functions include the following:

1. Improving the taxonomic classification system for Canadian soils and revision of this system because of new information.

2. Improving the identification of physical features and soil characteristics used in describing and mapping soils.

3. Reviewing the methods, techniques, and nomenclature used in soil surveys and the recommendation of changes necessary for either a greater measure of uniformity or for their improvement.

4. Recommending investigations of problems affecting soil classification, soil formation, and the interpretation of soil survey information.

5. Recommending and supporting investigations on interpretations of soil survey information for soil ratings, crop yield assessments, soil mechanics, and other purposes.

6. Cooperating with specialists in soil fertility, agronomy, agrometeorology, and other disciplines in assessing interrelated problems.

Much of the credit for the present degree of realization of these objectives is due to A. Leahey, chairman from 1940 to 1966, and P.C. Stobbe, secretary from 1940 to 1969. W.A. Ehrlich was chairman from 1966 to 1971 and was succeeded by J.S. Clark. In 1969 the name of the committee was changed to the Canada Soil Survey Committee (CSSC).

Developments in soil classification in Canada since 1940 are documented in reports of the meetings of the NSSC held in 1945, 1948, 1955, 1960, 1963, 1965, and 1968 and of the CSSC in 1970, 1973, 1974, and 1976. Soil classification was one of the main items on the agenda of the first meeting, and a report by P.C. Stobbe provoked a prolonged and lively discussion. He and his committee recommended a system of field classification similar to that developed in Manitoba by Ellis (1932). The proposed system was a hierarchical one with seven categories as follows:

Soil Regions—tundra, woodland, and grassland soils.

Soil Zones—broad belts in which a dominant kind of soil occurs, such as podzol or black soil.

Soil Subzones—major subdivisions of soil zones, such as black and degraded black.

Soil Associations or Catenas—the group of soils that occur together on the same parent material to form a land pattern.

Soil Series, Members, or Associates—the individual kinds of soils that are included in an association.

Soil Type or Soil Class—subdivisions of associations or of series based upon texture.

Soil Phase—subdivisions of mapping units based upon external soil characteristics such as stoniness and topography.

This proposal was for a field system of classification, or a system for classifying the units of soils mapped at various scales. The classes at all levels, phase to region, were segments of the landscape that included all the soil variability within the area designated. Thus, a soil zone was a land area in which a "zonal great soil group occurred as a dominant soil." The system was not intended to be a scientific or taxonomic one in which the classes at all levels had clearly defined limits based on a reasonably thorough knowledge of the properties of the entire population of soils in Canada. The proposed system was accepted for trial by the committee, which represented all provinces. Thus, an important step was taken in the development of a national system of classifying the units of soil mapped in soil surveys.

The first Canadian taxonomic system of soil classification was presented by P.C. Stobbe at the NSSC meeting in 1955. This system was a marked departure from the mapping or field classification system proposed in 1945. It

probably resulted from the following circumstances:

- the greater knowledge of Canadian soils
- the desire to classify soils, even at the highest categorical level, based on the properties of the soils themselves
- the need for a taxonomic system better than the old USDA system (Baldwin et al. 1938) that focused unduly on "normal" soils. The Soil Conservation Service had begun to develop a new system in 1951, but the fourth approximation of that system was judged to be too complicated and too tentative for Canadian needs (National Soil Survey Committee 1955).

Unfortunately, formal discussion of field systems of classifying soils or soil mapping systems was dropped for several years at NSSC meetings, but the need for such systems was recognized by leaders in pedology. This need can be illustrated by an example of mapping soils at a particular scale and classifying the kinds of soil that occur.

If the map is at a scale of 1:100 000 and the smallest area delineated is a square measuring 1 cm on each side, that area represents 100 ha. Such an area commonly includes upland and lowland positions in the landscape and the associated kinds of soils. The kinds of soils within that area are identified by digging pits at different topographic positions in the landscape. At each of these points the profile exposed usually has a rather narrow range of properties that reflects the influence of soil-forming factors at that point. Therefore, the soils at each point of observation can be classified as a single class in a taxonomic system. The area delineated on the map cannot be classified as a single class in such a system because it includes several kinds of soils. However, the area mapped can be classified as a kind of soil mapping unit such as a soil association in the system of Ellis (1932). Thus, the need was evident for both a taxonomic system to permit the naming and the ordering of information about specific kinds of soils, and a mapping system to permit the ordering of information about the areas delineated on soil maps and the naming of them.

The taxonomic system outlined in 1955, which is the basis of the system used today, had six unnamed categorical levels corresponding to the order, great group, subgroup, family, series, and type. The seven taxa separated at the order level were: Chernozemic,

Halomorphic, Podzolic, Forested Brown, Regosolic, Gleisolic, and Organic. Taxa were defined only in general terms down to the subgroup level. Although this was inevitable because of the lack of sufficient information, it led to differences of interpretation of the taxa in various provinces and some lack of uniformity in the use of the system. The need for correlation was clearly recognized by senior Canadian pedologists.

Progress in developing the Canadian system of soil classification since 1955 has been toward more precisely defining the taxa at all categorical levels and increasingly emphasizing soil properties as taxonomic criteria. This is evident from the reports of NSSC meetings held in 1960, 1963, 1965, and 1968, at which the main topic of discussion was soil classification. Some changes in taxa at the order, great group, and subgroup levels were made at these meetings. For example, in 1963 the Meadow and Dark Gray Gleisolic great groups were combined as Humic Gleysol; in 1965 a system of classifying soils of the Organic order was presented and accepted; in 1968 the former Podzolic order was divided into Luvisolic (clay translocation) and Podzolic (accumulation of Al and Fe organic complexes) orders, and the concept and classification of Brunisolic soils were revised. Criteria of classification involving morphological, chemical, and physical properties became increasingly specific through this period. The bases of classifying soils at the family level were outlined and the series and type categories were defined more specifically.

Following the publication of *The System of Soil Classification for Canada* in 1970, topics other than soil taxonomy were emphasized at CSSC meetings. However, in 1973 a Cryosolic order was proposed to classify the soils with permafrost close to the surface, and some refinements were made in several orders. These changes, including the newly developed Cryosolic order, were incorporated into *The Canadian System of Soil Classification* published in 1978.

Between 1945 and 1970 little consideration was given at NSSC meetings to systems of naming and classifying soil mapping units. Between 1970 and 1978 a satisfactory taxonomic system was developed. In 1978 the Expert Committee on Soil Survey was formed, replacing the Canada Soil Survey Committee. Although the work of the Canada Soil Survey Committee focused on interpretation, mapping systems, and soil degradation, soil classification also formed an important part of its activities.

In 1980, after C. Tarnocai became chairperson of the Soil Classification Working Group, work began on a number of problems in soil classification that had been identified by soil scientists from various regions. The main items covered included the classification of Gleysols (McKeague et al. 1986), Folisols (Fox 1985, Fox et al. 1987, Trowbridge 1981, and Trowbridge et al. 1985), Podzols and the definition of contrasting horizons. At a meeting of the working group held in 1984, proposals were presented on these topics and solutions were formulated and presented to the Expert Committee on Soil Survey (Tarnocai 1985). The Second Edition of *The Canadian System of Soil Classification* published in 1987 incorporated the recommendations accepted by the committee.

In the late 1980s, as a result of an international soil correlation tour organized by the USDA Soil Conservation Service, attention was drawn to Vertisolic soils; soils of high clay content whose properties result from shrinking and swelling, caused by wetting and drying. Previously, Vertisolic soils were thought to occur only in temperate and tropical climates. However, when international experts attending this meeting examined high clay content soils in southern Saskatchewan, they found them to have the same properties as Vertisols occurring elsewhere and considered these soils to be Vertisols. This changed viewpoint initiated activities aimed at incorporating Vertisols in *The Canadian System of Soil Classification*. The Canadian Soil Classification Working Group formulated a number of proposals for classifying these soils. Between 1991 and 1994 field tours were held in the Prairie Provinces, British Columbia, and eastern Canada to test these proposals in the field. As a result of these activities, the working group recommended establishing a Vertisolic order to classify these soils and set up criteria for this new soil order.

This Third Edition of *The Canadian System of Soil Classification* includes minor revisions throughout the system and introduces the Vertisolic order. This edition also introduces the Vertic great group for Solonetzic soils and Vertic subgroups for the Chernozemic, Gleysolic, Luvisolic, and Solonetzic soils. It also provides a major revision of the Cryosolic order resulting from international activities, both in the United States and Russia.

Rationale of Soil Taxonomy in Canada

During some 80 years of pedological work in Canada, concepts of soil and systems of classification have progressed as a result of new knowledge and new concepts developed in Canada and elsewhere. Here an attempt is made to enunciate the current rationale of soil taxonomy based on the historical material outlined in the previous section and on recent publications on pedology in Canada.

The nature of soil

The concept of soil in Canada and elsewhere (Cline 1961; Knox 1965; Simonson 1968) has changed greatly since 1914 when the first soil survey was started in Ontario. No specific definition is available from that early work, but clearly soil was thought of as the uppermost geological material. Texture was apparently considered to be its most important attribute. Currently, soil is defined in general terms by pedologists as the naturally occurring, unconsolidated, mineral or organic material at the earth's surface that is capable of supporting plant growth. Its properties usually vary with depth. They are determined by climatic factors and organisms, as conditioned by relief and, hence, water regime acting on geological materials and producing genetic horizons that differ from the parent material. In the landscape, soil merges into nonsoil entities such as exposed, consolidated rock or permanent bodies of water at arbitrarily defined boundaries. Specific definitions of soil and nonsoil are given in Chapter 2.

Because soil occurs at the surface of the earth as a continuum with variable properties, it is necessary to decide on a basic unit of soil to be described, sampled, analyzed, and classified. Such a unit was defined by United States' pedologists (Soil Survey Staff 1960) and is accepted in Canada. Called a pedon, it is the smallest, three-dimensional body at the surface of the earth that is considered to be a soil. Its lateral dimensions are 1–3.5 m and its depth is 1–2 m. The pedon is defined more specifically in Chapter 2.

Nature and purpose of soil classification

Soil classification systems are not truths that can be discovered but rather are methods of organizing information and ideas in ways that seem logical and useful (Soil Survey Staff 1960). Thus no classification system is either true or false; some systems are more logical and useful for certain objectives than others. A classification system reflects the existing knowledge and concepts concerning the population of soils being classified (Cline 1949). It must be modified as knowledge grows and new concepts develop.

Both the theoretical and practical purposes of soil classification have been discussed in the literature (Cline 1949, 1963; De Bakker 1970). The general purpose of soil classification in Canada may be stated as follows:

- to organize the knowledge of soils so that it can be recalled systematically and communicated, and that relationships may be seen among kinds of soils, among soil properties and environmental factors, and among soil properties and suitabilities of soils for various uses.

The related purposes of soil classification are to provide a framework for formulating hypotheses about soil genesis and the response of soil to management, to aid in extending knowledge of soils gained in one area to other areas having similar soils, and to provide a basis for indicating the kinds of soils within mapping units. Soil classification is essential to soil surveys, to the teaching of soils as a part of natural science, and to meeting the practical needs related to land use and management.

The overall philosophy of the Canadian system is pragmatic; the aim is to organize the knowledge of soils in a reasonable and usable way. The system is a natural, or taxonomic, one in which the classes (taxa) are based upon properties of the soils themselves and not upon interpretations of the soils for various uses. Interpretations involve a second step that is essential if the information is to be used effectively. If the taxa are defined on the basis of soil properties, and if the boundaries of these classes, or of combinations of them, are shown on a map, interpretations can be made on the basis of properties implied in the class definitions.

Misconceptions about soil taxonomy

Misconceptions about the functions of a system of soil taxonomy are evident periodically. Some of these are listed to warn users of the Canadian system against unrealistic expectations.

1. It is a misconception that a good system results in the assignment of soils occurring close together to the same taxon, at least the higher categorical levels. This is neither possible, nor desireable, in some areas. Pedons a few metres apart may differ as greatly as pedons hundreds of kilometres apart within a climatic region.

2. Another common misconception is that a good national system provides the most suitable groupings of soils in all areas. This is not possible because criteria based upon properties of the whole population of soils in the country are bound to differ from those developed on the basis of properties of soils in any one region. Criteria developed for a national system will inevitably result in areas where most of the soils have properties that straddle the boundary line between two taxa.

3. The idea that if the system was soundly based there would be no need for changes every few years is erroneous. As new areas are surveyed, as more research is done, and as concepts of soil develop, changes in the system become inevitable to maintain a workable taxonomy.

4. Another unfortunate hope is that a good system will ensure that taxa at the order level at least can be assigned unambiguously and easily in the field. Actually, in a hierarchical system the divisions between orders must be defined just as precisely as those between series. With pedons having properties close to class boundaries at any taxonomic level, classification is difficult and laboratory data may be necessary.

5. The assumption is made by some that a good system permits soils occurring within mapped areas to be classified as members of not more than three series. Clearly, this is not reasonable because the number of taxononic classes occurring within a mapping unit depends upon the complexity of the pattern of soils in the landscape, on the scale of the map, and on the narrowness of class limits.

6. The idea that a good system is simple enough to be clear to any layman is erroneous. Unfortunately, soil is complex and, although the general ideas of the taxonomy should be explainable in simple terms, the definitions of taxa must be complex in some instances.

7. Another misconception is that a good system makes soil mapping easy. Ease of mapping depends more upon the complexity of the landscape, the access, and the predictability of the pattern of soils within segments of the landscape than upon taxonomy.

Attributes of the Canadian system

The development of soil taxonomy in Canada has been toward a system with the following attributes:

1. It provides taxa for all known soils in Canada.

2. It involves a hierarchical organization of several categories to permit soils at various levels of generality to be considered. Classes at high categorical levels reflect, to the extent possible, broad differences in soil environments that are related to differences in soil genesis.

3. The taxa are defined specifically so as to convey the same meaning to all users.

4. The taxa are concepts based upon generalizations of properties of real bodies of soils rather than idealized concepts of the kinds of soils that would result from the action of presumed genetic processes. The criteria chosen define taxa in accordance with desired groupings of soils. The groupings are not decided upon initially on the basis of arbitrary criteria.

5. Differentiae among the taxa are based upon soil properties that can be observed and measured objectively in the field or, if necessary, in the laboratory.

6. The system can be modified on the basis of new information and concepts without destroying the overall framework. Periodically, however, the entire framework of the system will be reevaluated.

Although taxa in the Canadian system are defined on the basis of soil properties, the system has a genetic bias in that properties or combinations of properties that reflect genesis are favored as differentiae in the higher categories. For example, the use of the terms chernozemic A and podzolic B imply genesis. The reason for the genetic bias is that it seems reasonable to combine, at high categorical levels, soils in which particular horizons developed under similar dominant processes resulting from broadly similar climatic conditions. Classification is not based directly on presumed genesis because soil genesis is incompletely understood, is subject to a wide variety of opinion, and cannot be measured simply.

Bases of criteria for defining taxa at various categorical levels

The bases for differentiating taxa at the various categorical levels are not clear cut. In a hierarchical system of soil classification, logical groupings of soils that reflect environmental factors cannot be obtained by following any rigid systematic framework in which all taxa at the same categorical level are differentiated based on a uniform specific criterion, such as acidity or texture. The fact that criteria must be based on soil properties, rather than directly on environmental factors or use evaluation was recognized by some pedologists almost three-quarters of a century ago (Joel 1926). The general bases of the different categorical levels, which are presented below, can be inferred from a study of the system. They apply better to some taxa than to others; for example, the statement for order applies more clearly to Chernozemic and Podzolic soils than to Regosolic and Brunisolic soils.

Order Taxa at the order level are based on properties of the pedon that reflect the nature of the soil environment and the effects of the dominant, soil-forming processes.

Great group Great groups are soil taxa formed by subdividing each order. Thus, each great group carries with it the differentiating criteria of the order to which it belongs. In addition, taxa at the great group level are based on properties that reflect differences in the strengths of dominant processes, or a major contribution of a process in addition to the dominant one. For example, in Luvic Gleysols the dominant process is considered to be gleying, but clay translocation is also a major process.

Subgroup Subgroups are formed by subdividing each great group. Therefore, they carry the differentiating criteria of the order and the great group to which they belong. Subgroups are also differentiated on the basis of the kind and arrangement of horizons that indicate conformity to the central concept of the great group (e.g., Orthic), intergrading toward soils of another order (e.g., Gleyed, Brunisolic), or additional special features within the control section (e.g., Ortstein, Vertic).

Family Taxa at the family level are formed by subdividing subgroups. Thus, they carry the differentiating criteria of the order, great group, and subgroup to which they belong. Families within a subgroup are differentiated based on parent material characteristics, such as particle size, mineralogy, calcareousness, reaction, and depth, and on soil climatic factors.

Series Series are formed by subdividing families. Therefore, they carry all the differentiating criteria of the order, great group, subgroup, and family to which they belong. Series within a family are differentiated on the basis of detailed features of the pedon. Pedons belonging to a series have similar kinds and arrangements of horizons whose color, texture, structure, consistence, thickness, reaction, and composition fall within a narrow range. A series is a category in the taxonomic system; thus, it is a conceptual class in the same sense as an order.

A pedon is a real unit of soil in the landscape; a series is a conceptual class with defined limits based on the generalization of properties of many pedons. A particular pedon may be classified as a series if its attributes fall within the limits of those of an established series. However, it is not, strictly speaking, a series because the attributes of any one pedon do not encompass the complete range of attributes allowable within a series. Thus, it is not correct to study part of a pedon and to declare, "this is X series." Rather it should be stated, "this pedon has properties that fall within the limits of the X series," or "this pedon is classified in the X series."

Relationship of taxonomic classes to environments

A general relationship exists between kinds of environments and taxa at various levels in the system. This follows from the basis of selecting diagnostic criteria for the taxa; the primary basis at the higher levels is to select properties that reflect the environment and properties resulting from processes of soil genesis. Although the system may look like a key with classes defined precisely but arbitrarily on the basis of specific properties, it is one in which the taxa reflect, to as great an extent as possible, genetic or environmental factors.

The Podzolic order, for example, is defined based on morphological and chemical properties of the B horizon. However, these properties are associated with humid conditions, sandy to loamy parent materials, and forest or heath vegetation. Although the great groups within the order are defined on the basis of the amounts of organic C and extractable Fe and Al in the B horizons, they have broad environmental significance. Humic Podzols are associated with very wet environments, high water tables, periodic or continuous reducing conditions, hydrophytic vegetation, and, commonly, a peaty surface. Ferro-Humic Podzols occur in areas of high effective precipitation, but they are not under reducing conditions for prolonged periods. Humo-Ferric Podzols generally occur in less humid environments than the other great groups in the order. An interrelation of climatic and vegetative factors, parent material, and relief determines the occurrence of the different classes of Podzolic soils. Similarly, general relationships exist between other orders, great groups, and soil environmental factors. However, these relationships are much less clear for some Regosolic and Brunisolic soils than they are for most soils of other orders. At lower categorical levels, in general, relationships between soil taxa and factors of the soil environment become increasingly close.

Relationship of the Canadian system to other systems of soil taxonomy

The numerous national systems of soil taxonomy might be looked upon as indications of the youthfulness of soil science. Knowledge of the properties of the soils of the world is far from complete, therefore it is not possible to develop an international system of classification for the whole population of known and unknown soils. Probably even after such a system has been developed, national systems will remain in use because they are familiar and are thought to be more useful for the restricted population of soils within the country. The soil units defined for the FAO-UNESCO world soil map project are useful in international soil correlation, but they do not constitute a complete system of soil taxonomy (FAO 1985). The closest approach to a comprehensive system of soil taxonomy is that produced by the Soil Survey Staff of USDA (1994), which has been under development since 1951. Like previous U.S. systems, it has had a major influence on soil taxonomy in Canada and elsewhere.

The Canadian system of soil taxonomy is more closely related to the U.S. system than to any other. Both are hierarchical, and the taxa are defined on the basis of measurable soil properties. However, they differ in several respects. The Canadian system is designed to classify only soils that occur in Canada and is not a comprehensive system. The U.S. system has a suborder, which is a category that the Canadian system does not have. In the Canadian system Solonetzic and Gleysolic soils are differentiated at the highest categorical level as in the Russian and some other European systems. These soils are differentiated at the suborder or great group level in the U.S. system. Perhaps the main difference between the two systems is that all horizons to the surface may be diagnostic in the Canadian system, whereas horizons below the depth of plowing are emphasized in the U.S. system. This may be because 90% of the area of Canada is not likely to be cultivated.

Summary

The Canadian system is a hierarchical one in which the classes are conceptual, based upon the generalization of properties of real bodies of soil. Taxa are defined on the basis of observable and measurable soil properties that reflect processes of soil genesis and environmental factors. The development of the system has progressed with the increasing knowledge of the soils of Canada obtained through pedological surveys carried out over an 80-year period. The system has been influenced strongly by concepts developed in other countries, but some aspects are uniquely Canadian. The system is imperfect because it is based on a limited knowledge of the vast population of soils in the country. However, the system does make it possible to assign soils throughout Canada to taxa at various levels of generalization and to organize the knowledge of soils in such a way that relationships between factors of the environment and soil development can be seen. It is possible to define the kinds of soils that occur within units on soil maps, and to provide a basis for evaluating mapped areas of soil for a variety of potential uses.

Chapter 2

Soil, Pedon, Control Section, and Soil Horizons

Soil taxonomy in Canada is based on properties of the soil. Therefore, there is a need to define soil (as opposed to nonsoil) and the unit of soil being classified. These and the equally basic definitions of soil horizons are given in this chapter.

Soil and Nonsoil

Bodies of soil and nonsoil occur as a continuum at the surface of the earth. They merge into one another often at imperceptible though arbitrarily defined boundaries. Soil is defined herein as the naturally occurring, unconsolidated mineral or organic material at least 10 cm thick that occurs at the earth's surface and is capable of supporting plant growth. In this definition "naturally occurring" includes disturbance of the surface by activities of man such as cultivation and logging but not displaced materials such as gravel dumps and mine spoils. Unconsolidated material includes material compacted or cemented by soil-forming processes. Soil extends from the earth's surface through the genetic horizons, if present, into the underlying material to the depth of the control section. Soil may have water covering its surface to a depth of 60 cm or less either at low tide in coastal areas or during the driest part of the year in areas inland. A soil covered by a veneer of new material at least 50 cm thick is considered to be a buried soil. Soil development involves climatic factors and organisms, as conditioned by relief and hence water regime, acting through time on geological materials and thus modifying the properties of the parent material.

Nonsoil is the aggregate of surficial materials that do not meet the preceding definition of soil. It includes soil materials displaced by unnatural processes such as dumps of earth fill, unconsolidated mineral or organic material thinner than 10 cm overlying bedrock, exposed bedrock, and unconsolidated material covered by more than 60 cm of water throughout the year.

Nonsoil also includes organic material thinner than 40 cm overlying water.

The definitions reflect the fact that bodies of soil and nonsoil have a continuum of properties. For example, the thickness of soil material overlying bedrock might range from 1 m at the base of a slope to 20 cm at midslope and gradually thin out to exposed bedrock at the top. The exposed bedrock is nonsoil, but the thickness of unconsolidated material over bedrock that should qualify as soil is not obvious. To avoid ambiguity and permit uniformity of classification, an arbitrary depth limit of 10 cm is used. Similarly, bodies of periodically submerged soil merge into bodies of water in the natural landscape.

Pedon, the Basic Unit of Soil

The pedon as defined by pedologists of the United States (Soil Survey Staff 1975) serves as the basic unit of soil in the Canadian classification system. It is the smallest, three-dimensional unit at the surface of the earth that is considered as a soil. Its lateral dimensions are 1 m if ordered variation in genetic horizons can be sampled within that distance or if these horizons are few and faintly expressed. If horizons are cyclical or intermittent and are repeated in a lateral distance of 2–7 m, the lateral dimensions of the pedon are half the cycle (1–3.5 m). The vertical dimension of the pedon is to the depth of the control section. A contiguous group of similar pedons is called a polypedon, which is indicated in *Soil Taxonomy* (Soil Survey Staff 1975) as a unit of classification.

The pedon concept applies to the classification of all soils, but its relevance to soils having cyclic variation, such as Turbic Cryosols, is most apparent. Examples of pedons are illustrated (Figs. 1, 2, and 3). In Figure 1 the profiles beneath the nonsorted circles and the intercircle material differ markedly. However, the variation is cyclic

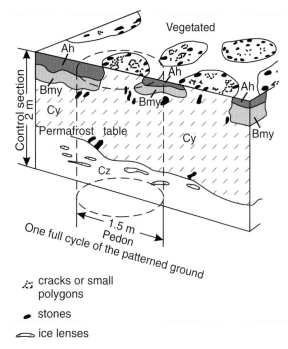

Figure 1 Pedon of Orthic Turbic Cryosol in area of nonsorted circles.

and occurs repeatedly over the landscape. In the case of circles, the patterned ground unit is delimited by the trough that forms the perimeter of the circle. The diameter of the circle is measured from the midpoint of the trough on one side of the circle to its midpoint on the other side. If this diameter is no greater than 2 m, the full cycle forms the pedon (Fig. 1). All the variability within the pedon is included in the concept of the taxonomic class used from order to series. In this case classification would be based upon the properties of the intercircle material as it is dominant in extent. If the circles were further apart such that a full cycle was 2–7 m in diameter, the pedon would include half a cycle. Thus it would extend from the midpoint of a circle to the midpoint of the trough or intercircle material. If the circles were still further apart, such that the lateral dimension of the cycle was greater than 7 m, two pedons would be identified. Since this is common with ice-wedge polygons, one pedon associated with the ice-wedge polygon trench or trough and the other associated with the central part of the ice-wedge polygon or circle would be identified.

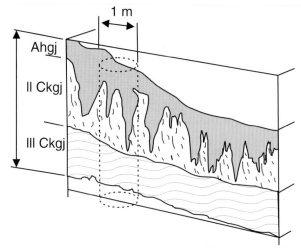

Figure 2 Pedon of Gleyed Vertic Black Chernozem with tonguing Ah horizon.

Cyclic variation involving tonguing of Ah horizon material into the IICk horizon (see Soil Horizons and Other Layers) is shown in Figure 2. The full cycle has lateral dimensions of about 1 m or less so that the pedon includes a full cycle. All the variability in thickness of the Ah is included in the concept of the soil series.

The pedon is half of the cycle in the example of hummocky terrain resulting from the blowdown of trees shown in Figure 3. In such cases the hummocks are usually not regularly distributed and the dimensions of the pedon may vary appreciably over short distances.

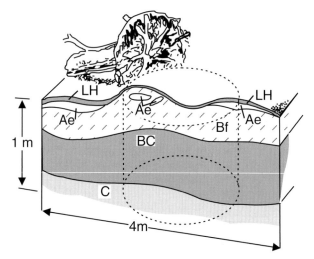

Figure 3 Pedon of Orthic Humo-Ferric Podzol, turbic phase, in hummocky terrain due to blowdown of trees.

Control Section

The control section is the vertical section of soil upon which classification is based. It is necessary to provide a uniform basis for soil classification. In general, pedons should be sampled at least to the depth of the control section. The properties of the material beneath the control section are important for many interpretive purposes. Therefore, the underlying material should be examined and its properties recorded whenever possible.

Mineral soils

For mineral soils in general, the control section extends from the mineral surface either to 25 cm below the upper boundary of the C, IIC, or permafrost table, or to a depth of 2 m, whichever is less. Exceptions are as follows:

1. If the upper boundary of the C or IIC is less than 75 cm from the mineral surface, the control section extends to a depth of 1 m.
2. If bedrock occurs at a depth of 10 cm or more but less than 1 m, the control section extends from the surface to the lithic contact.
3. If permafrost occurs at a depth of less than 1 m and the soil does not show evidence of cryoturbation (Static Cryosol), the control section extends to a depth of 1 m.

Organic soils

A. Organic Order

The control section for Fibrisols, Mesisols, and Humisols extends from the surface either to a depth of 1.6 m or to a lithic contact. It is divided into tiers, which are used in classification. The tiers are layers based upon arbitrary depth criteria.

Surface tier The surface tier is 40 cm thick exclusive of loose litter, crowns of sedges and reeds, or living mosses. Mineral soil on the surface of the profile is part of the surface tier, which is used to name the soil family. Shallow lithic organic soils may have only a surface tier.

Middle tier The middle tier is 80 cm thick. It establishes the great group classification if no terric, lithic, or hydric substratum is present. Otherwise the dominant kind of organic material in this and the surface tier establishes the great group classification. The nature of the subdominant organic material in the middle or bottom tier assists in establishing the subgroup classification.

Bottom tier The bottom tier is 40 cm thick. The material in this tier establishes in whole or in part the subgroup classification.

The control section for Folisols is the same as that used for mineral soils. These soils must have more than 40 cm of folic materials if they overlie mineral soils or peat materials, or at least 10 cm if they overlie a lithic contact or fragmental materials.

B. Organic Cryosol Great Group

The control section for Organic Cryosols extends to a depth of 1 m or to a lithic contact. No tiers are defined.

Soil Horizons and Other Layers

The definitions of taxa in the Canadian system are based mainly on the kinds, degree of development, and the sequence of soil horizons and other layers in pedons. Therefore, the clear definition and designation of soil horizons and other layers are basic to soil classification. A soil horizon is a layer of mineral or organic soil material approximately parallel to the land surface that has characteristics altered by processes of soil formation. It differs from adjacent horizons in properties such as color, structure, texture, and consistence and in chemical, biological, or mineralogical composition. The other layers are either nonsoil layers such as rock and water or layers of unconsolidated material considered to be unaffected by soil-forming processes. For the sake of brevity these other layers are referred to simply as layers but it is recognized that soil horizons are also layers. In previous editions of this publication and in the *Glossary of Terms in Soil Science* (Canada Department of Agriculture 1976) organic materials are designated as layers and not horizons.

The major mineral horizons are A, B, and C. The major organic horizons are L, F, and H, which are mainly forest litter at various stages of decomposition, and O, which is derived mainly from wetland vegetation. Subdivisions of horizons are labeled by adding lower-case suffixes to some of the major horizon symbols as with Ah or Ae. Well-developed horizons are readily identified in the field. However, in cases of weak expression or of borderline properties, as between Ah and H, laboratory determinations are necessary before horizons can be designated positively. Many of the laboratory methods required are outlined in a publication sponsored by the Canadian Society of Soil Science (Carter 1993). Some other methods pertaining to organic horizons are outlined near the end of this chapter.

The layers defined are R, rock; W, water; and IIC or other nonconforming, unconsolidated mineral layers, IIIC, etc. below the control section that are unaffected by soil-forming processes. Theoretically a IIC affected by soil-forming processes is a horizon; for example a IICca is a horizon. In practice, it is usually difficult to determine the lower boundary of soil material affected by soil-forming processes. Thus the following are considered as horizons: C(IC), any unconforming layer within the control section, and any unconforming layer below the control section that has been affected by pedogenic processes (e.g., IIBc, IIIBtj). Unconforming layers below the control section that do not appear to have been affected by pedogenic processes are considered as layers. The tiers of Organic soils are also considered as layers.

Mineral horizons and layers

Mineral horizons contain 17% or less organic C (about 30% organic matter) by weight.

A—This mineral horizon forms at or near the surface in the zone of leaching or eluviation of materials in solution or suspension, or of maximum in situ accumulation of organic matter or both. The accumulated organic matter is usually expressed morphologically by a darkening of the surface soil (Ah). Conversely, the removal of organic matter is usually expressed by a lightening of the soil color usually in the upper part of the solum (Ae). The removal of clay from the upper part of the solum (Ae) is expressed by a coarser soil texture relative to the underlying subsoil layers. The removal of iron is indicated usually by a paler or less red soil color in the upper part of the solum (Ae) relative to the lower part of the subsoil.

B—This mineral horizon is characterized by enrichment in organic matter, sesquioxides, or clay; or by the development of soil structure; or by a change of color denoting hydrolysis, reduction, or oxidation. In B horizons, accumulated organic matter (Bh) is evidenced usually by dark colors relative to the C horizon. Clay accumulation is indicated by finer soil textures and by clay cutans coating peds and lining pores (Bt). Soil structure developed in B horizons includes prismatic or columnar units with coatings or stainings and significant amounts of exchangeable sodium (Bn) and other changes of structure (Bm) from that of the parent material. Color changes include relatively uniform browning due to oxidation of iron (Bm), and mottling and gleying of structurally altered material associated with periodic reduction (Bg).

C—This mineral horizon is comparatively unaffected by the pedogenic processes operating in A and B horizons, except the process of gleying (Cg), and the accumulation of calcium and magnesium carbonates (Cca) and more soluble salts (Cs, Csa). Marl, diatomaceous earth, and rock with a hardness ≤ 3 on Mohs' scale are considered to be C horizons.

R—This consolidated bedrock layer is too hard to break with the hands (>3 on Mohs' scale) or to dig with a spade when moist. It does not meet the requirements of a C horizon. The boundary between the R layer and any overlying unconsolidated material is called a lithic contact.

W—This layer of water may occur in Gleysolic, Organic, or Cryosolic soils. Hydric layers in Organic soils are a kind of W layer as is segregated ice formation in Cryosolic soils.

Lowercase suffixes

b—A buried soil horizon.

c—A cemented (irreversible) pedogenic horizon. Ortstein, placic and duric horizons of Podzolic soils, and a layer cemented by $CaCO_3$ are examples.

ca—A horizon of secondary carbonate enrichment in which the concentration of lime exceeds that in the unenriched parent material. It is more than 10 cm thick, and its $CaCO_3$ equivalent exceeds that of the parent material by at least 5% if the $CaCO_3$ equivalent is less than 15% (13% vs 8%), or by at least 1/3 if the $CaCO_3$ equivalent of the horizon is 15% or more (28% vs 21%). If no IC is present, this horizon is more than 10 cm thick and contains more than 5% by volume of secondary carbonates in concretions or in soft, powdery forms.

cc—Cemented (irreversible) pedogenic concretions.

e—A horizon characterized by the eluviation of clay, Fe, Al, or organic matter alone or in combination. When dry, it is usually higher in color value by one or more units than an underlying B horizon. It is used with A (Ae or Ahe).

f—A horizon enriched with amorphous material, principally Al and Fe combined with organic matter. It must have a hue of 7.5YR or redder, or its hue must be 10YR near the upper boundary and become yellower with depth. When moist the chroma is higher than 3 or the value is 3 or less. It contains at least 0.6% pyrophosphate-extractable Al+Fe in textures finer than sand and 0.4% or more in sands (coarse sand, sand, fine sand, and very fine sand). The ratio of pyrophosphate-extractable Al+Fe to clay (≤ 0.002 mm) is more than 0.05 and organic C exceeds 0.5 %. Pyrophosphate-extractable Fe is at least 0.3%, or the ratio of organic C to pyrophosphate-extractable Fe is less than 20, or both are true. It is used with B alone (Bf), with B and h (Bhf), with B and g (Bfg), and with other suffixes. These criteria do not apply to Bgf horizons. The following f horizons are differentiated on the basis of the organic C content:

Bf—0.5–5% organic C.

Bhf—more than 5% organic C.

No minimum thickness is specified for a Bf or a Bhf horizon. Thin Bf and Bhf horizons do not qualify as podzolic B horizons as defined later in this chapter. Some Ah and Ap horizons contain sufficient pyrophosphate-extractable Al+Fe to satisfy this criterion of f but are designated Ah or Ap.

g—A horizon characterized by gray colors, or prominent mottling, or both, indicating permanent or periodic intense reduction. Chromas of the matrix are generally 1 or less. It is used with A and e (Aeg); B alone (Bg); B and f (Bfg, Bgf); B, h, and f (Bhfg); B and t (Btg); C alone (Cg); C and k (Ckg); and several others. In some reddish parent materials matrix colors of reddish hues and high chromas may persist despite long periods of reduction. In these soils, horizons are designated as g if there is gray mottling or marked bleaching on ped faces or along cracks.

Aeg—This horizon must meet the definitions of A, e, and g.

Bg—This horizon is analogous to a Bm horizon but has colors indicating poor drainage and periodic reduction. It includes horizons occurring between A and C horizons in which the main features are as follows:
(i) Colors of low chroma: that is, chromas of 1 or less, without mottles on ped surfaces or in the matrix if peds are lacking; or chromas of 2 or less in hues of 10YR or redder, on ped surfaces or in the matrix if peds are lacking, accompanied by more prominent mottles than those in the C horizon; or hues bluer than 10Y, with or without mottles on ped surfaces or in the matrix if peds are lacking.
(ii) Colors indicated in (i) and a change in structure from that of the C horizon.
(iii) Colors indicated in (i) and illuviation of clay too slight to meet the requirements of Bt, or an accumulation of iron oxide too slight to meet the limits of Bgf.
(iv) Colors indicated in (i) and the removal of carbonates. Bg horizons occur in some Orthic Humic Gleysols and some Orthic Gleysols.

Bfg, Bhfg, Btg, and others—When used in any of these combinations, the limits set for f, hf, t, and others must be met.

Bgf—The dithionite-extractable Fe of this horizon exceeds that of the IC by 1% or more. Pyrophosphate-extractable Al+Fe is less than the minimum limit specified for f horizons. This horizon occurs in Fera Gleysols and Fera Humic Gleysols and possibly below the Bfg of gleyed Podzols. It is distinguished from the Bfg of gleyed Podzols on the basis of the extractability of the Fe and Al. The Fe in the Bgf horizon is thought to have accumulated as a result of the oxidation of ferrous iron. The iron oxide formed is not associated intimately with organic matter or with Al and is sometimes crystalline. The Bgf horizons are usually prominently mottled; more than half of the soil material occurs as mottles of high chroma.

Cg, Ckg, Ccag, Csg, Csag—When g is used with C alone, or with C and one of the lowercase suffixes k, ca, s, or sa, the horizon must meet the definition for C and for the particular suffix as well as for g.

h—A horizon enriched with organic matter. It is used with A alone (Ah), or with A and e (Ahe), or with B alone (Bh), or with B and f (Bhf).

Ah—This A horizon, enriched with organic matter, has a color value at least one unit lower than the underlying horizon or 0.5% more organic C than the IC or both. It contains 17% or less organic C by weight.

Ahe—This Ah horizon has undergone eluviation as evidenced, under natural conditions, by streaks and splotches of different shades of gray and often by platy structure. It may be overlain by a dark-colored Ah and underlain by a light-colored Ae.

Bh—This horizon contains more than 1% organic C, less than 0.3% pyrophosphate-extractable Fe, and has a ratio of organic C to pyrophosphate-extractable Fe of 20 or more. Generally the color value and chroma are 3 or less when moist.

Bhf—Defined under f.

j—A modifier of suffixes e, f, g, n, t, and v. It is used to denote an expression of, but failure to meet, the specified limits of the suffix it modifies. It must be placed to the right and adjacent to the suffix it modifies. For example, Bfgj means a Bf horizon with a weak expression of gleying; Bfjgj means a B horizon with weak expression of both f and g features.

Aej—It denotes an eluvial horizon that is thin, discontinuous, or slightly discernible.

Btj—It is a horizon with some illuviation of clay but not enough to meet the limits of Bt.

Btgj, Bmgj—These horizons are mottled but do not meet the criteria of Bg.

Bfj—It is a horizon with some accumulation of pyrophosphate-extractable Al+Fe but not enough to meet the limits of Bf. In addition, the color of this horizon may not meet the color criteria set for Bf.

Btnj or Bnj—In these horizons the development of solonetzic B properties is evident but insufficient to meet the limits for Bn or Bnt.

Bvj—In this horizon argillipedoturbation is evident but the disruption of other horizons is insufficient to severely alter them.

k—Denotes the presence of carbonate as indicated by visible effervescence when dilute HCl is added. It is used mostly with B and m (Bmk) or C (Ck) and occasionally with Ah or Ap (Ahk, Apk), or organic layers (Ofk, Omk).

m—A horizon slightly altered by hydrolysis, oxidation, or solution, or all three to give a change in color or structure, or both. It has the following properties:

1. Evidence of alteration in one of the following forms:
   a. Higher chromas and redder hues than the underlying horizons.
   b. Removal of carbonates either partially (Bmk) or completely (Bm).
   c. A change in structure from that of the original material.
2. Illuviation, if evident, too slight to meet the requirements of a Bt or a podzolic B.
3. Some weatherable minerals.
4. No cementation or induration and lacks a brittle consistence when moist.

This suffix can be used as Bm, Bmj, Bmk, and Bms.

n—A horizon in which the ratio of exchangeable Ca to exchangeable Na is 10 or less. It must also have the following distinctive morphological characteristics: prismatic or columnar structure, dark coatings on ped surfaces, and hard to very hard consistence when dry. It is used with B as Bn or Bnt.

p—A horizon disturbed by man's activities such as cultivation, logging, and habitation. It is used with A and O.

s—A horizon with salts, including gypsum, which may be detected as crystals or veins, as surface crusts of salt crystals, by depressed crop growth, or by the presence of salt-tolerant plants. It is commonly used with C and k (Csk) but can be used with any horizon or combination of horizon and lowercase suffix.

sa—A horizon with secondary enrichment of salts more soluble than Ca and Mg carbonates; the concentration of salts exceeds that in the unenriched parent material. The horizon is at least 10 cm thick. The conductivity of the saturation extract must be at least 4 mS/cm and exceed that of the C horizon by at least one-third.

ss—Denotes the presence of several (more than two) slickensides. It is used with B alone (Bss), with B and other lower case suffixes (Bssk, Bssgj, Bsskgj, Btss, etc.), with C alone (Css), with C and other lower case suffixes (Ckss, Ckssgj, etc.), with AC (ACss) or with BC (BCss). Slickensides are shear surfaces, with an aerial extent of at least 4 cm^2, that form when one soil mass moves over another. They commonly display unidirectional grooves parallel to the direction of movement and often occur at an angle of 20–60 degrees from the horizontal. Slickensides often intersect, resulting in the formation of wedge shaped aggregates that commonly occur in these soils.

t—An illuvial horizon enriched with silicate clay. It is used with B alone (Bt), with B and g (Btg), with B and n (Bnt), etc.

Bt—A Bt horizon is one that contains illuvial layer lattice clays. It forms below an eluvial horizon but may occur at the surface of a soil that has been partially truncated. It usually has a higher ratio of fine clay to total clay than the IC. It has the following properties:

1. If any part of an eluvial horizon remains and there is no lithologic discontinuity between it and the Bt horizon, the Bt horizon contains more total clay than the eluvial horizon as follows:
a. If any part of the eluvial horizon has less than 15% total clay in the fine earth fraction (≤2 mm), the Bt horizon must contain at least 3% more clay (e.g., Ae 10% clay; Bt minimum 13% clay).
b. If the eluvial horizon has more than 15% and less than 40% total clay in the fine earth fraction, the ratio of the clay in the Bt horizon to that in the eluvial horizon must be 1.2 or more (e.g., Ae 25% clay; Bt at least 30% clay).
c. If the eluvial horizon has more than 40% total clay in the fine earth fraction, the Bt horizon must contain at least 8% more clay (e.g., Ae 50% clay; Bt at least 58% clay).

2. A Bt horizon must be at least 5 cm thick. In some sandy soils where clay accumulation occurs in the lamellae, the total thickness of the lamellae should be more than 10 cm in the upper 1.5 m of the profile.

3. In massive soils the Bt horizon should have oriented clay in some pores and also as bridges between the sand grains.

4. If peds are present, a Bt horizon has clay skins on some of the vertical and horizontal ped surfaces and in the fine pores or has illuvial oriented clays in 1% or more of the cross section as viewed in thin section.

5. If a soil shows a lithologic discontinuity between the eluvial horizon and the Bt horizon, or if only a plow layer overlies the Bt horizon, the Bt horizon need show only clay skins in some part, either in some fine pores or on some vertical and horizontal ped surfaces. Thin sections should show that the horizon has about 1% or more of oriented clay bodies.

Btj and **Btg** - Defined under j and g.

u—A horizon that is markedly disrupted by physical or faunal processes other than cryoturbation or argillipedoturbation caused by Vertisolic processes. Evidence of marked disruption such as the inclusion of material from other horizons or the absence of the horizon must be evident in at least half of the cross section of the pedon. Such turbation can result from a blowdown of trees, mass movement of soil on slopes, and burrowing animals.

The u can be used with any horizon or subhorizon with the exception of A or B alone; e.g., Aeu, Bfu, Bcu.

v—A horizon affected by argillipedoturbation, as manifested by disruption and mixing caused by shrinking and swelling of the soil mass. It is characterized by the presence of the following:

1. Irregular shaped, randomly oriented, intrusions of displaced materials within the solum.

2. Vertical cracks, often containing sloughed-in surface materials. The disruption within this horizon is strong enough to prevent the development of horizons diagnostic of other orders, or if these horizons are present they are disrupted to the extent that they are no longer continuous and their orientation has been severely changed. It is used with B or BC horizons alone or in combination with other suffixes; e.g., Bv, Bvk, Bvg, Bvgj, BCvj, etc.

x—A horizon of fragipan character. See definition of fragipan.

y—A horizon affected by cryoturbation as manifested by disrupted and broken horizons, incorporation of materials from other horizons, and mechanical sorting in at least half of the cross section of the pedon. It is used with A, B, and C alone or in combination with other subscripts, e.g., Ahy, Ahgy, Bmy, Cy, Cgy, Cygj.

z—A frozen layer. It may be used with any horizon or layer, e.g., Ohz, Bmz, Cz, Wz.

Named diagnostic horizons and layers of mineral soils

Chernozemic A—This A horizon has all the following characteristics:

1. It is at least 10 cm thick or is thick and dark enough to provide 10 cm of surface material that meets the color criteria given in 2 and 3.

2. It has a color value darker than 5.5 dry and 3.5 moist, and its chroma is less than 3.5 moist.

3. It has a color value at least one Munsell unit darker than that of the IC horizon.

4. It contains 1–17% organic C and its C:N ratio is less than 17.

5. Characteristically it has neither massive structure and hard consistence nor single-grained structure, when dry.

6. It has a base saturation (neutral salt) of more than 80% and Ca is the dominant exchangeable cation.

7. It is restricted to soils having a mean annual soil temperature of 0°C or higher and a soil moisture regime subclass drier than humid. Usually chernozemic A horizons are associated with well to imperfectly drained soils having cold semiarid to subhumid soil climates.

Duric horizon—This strongly cemented horizon does not satisfy the criteria of a podzolic B horizon. Usually it has an abrupt upper boundary to an overlying podzolic B or to a Bm horizon and a diffuse lower boundary more than 50 cm below. Cementation is usually strongest near the upper boundary, which occurs commonly at a depth of 40–80 cm from the mineral surface. The color of the duric horizon usually differs little from that of the moderately coarse textured to coarse textured parent material, and the structure is usually massive or very coarse platy. Air-dry clods of duric horizons do not slake when immersed in water, and moist clods at least 3 cm thick usually cannot be broken in the hands.

Fragipan—A fragipan is a loamy subsurface horizon of high bulk density and very low organic matter content. When dry, it has a hard consistence and seems to be cemented. When moist, it has moderate to weak brittleness. It frequently has bleached fracture planes and is overlain by a friable B horizon. Air-dry clods of fragic horizons slake in water.

Ortstein—This strongly cemented horizon (Bhc, Bhfc, or Bfc) is at least 3 cm thick and occurs in more than one-third of the exposed face of the pedon. Ortstein horizons are generally reddish brown to very dark reddish brown.

Placic horizon—This horizon is a thin layer (commonly 5 mm or less in thickness) or a series of thin layers that are irregular or involuted, hard, impervious, often vitreous, and dark reddish brown to black. Placic horizons may be cemented by Fe, Al-organic complexes (Bhfc or Bfc), hydrated Fe oxides (Bgfc), or a mixture of Fe and Mn oxides.

Podzolic B horizon—This diagnostic horizon is defined by morphological and chemical properties.

Morphological

1. It is at least 10 cm thick.

2. The moist crushed color is either black, or the hue is 7.5YR or redder or 10YR near the upper boundary and becomes yellower with depth. The chroma is higher than 3 or the value is 3 or less.

3. The accumulation of amorphous material is indicated by brown to black coatings on some mineral grains or brown to black microaggregates. Also there is a silty feel when the material is rubbed wet, unless it is cemented.

Chemical

Two kinds of podzolic B horizons are differentiated chemically.

1. Very low Fe. Such a podzolic B horizon (Bh) must be at least 10 cm thick and have more than 1% organic C, less than 0.3% pyrophosphate-extractable Fe, and a ratio of organic C to pyrophosphate extractable Fe of 20 or more.

2. Contains appreciable Fe as well as Al. Such a podzolic B horizon (Bf or Bhf) must be at least 10 cm thick and have an organic C content of more than 0.5%. It contains 0.6% or more pyrophosphate-extractable Al+Fe in textures finer than sand and 0.4% or more in sands (coarse sand to very fine sand). The ratio of pyrophosphate-extractable Al+Fe to clay (<2 µm) is more than 0.05. Pyrophosphate-extractable Fe is at least 0.3%, or the ratio of organic C to pyrophosphate-extractable Fe is less than 20, or both are true.

Not all Bh, Bhf, and Bf horizons are podzolic B horizons because a podzolic B has a thickness requirement whereas Bh, Bhf, and Bf horizons do not.

Solonetzic B horizon—The term includes both Bn and Bnt horizons. These horizons have prismatic or columnar primary structure that breaks to blocky secondary structure; both structural units have hard to extremely hard consistence when dry. The ratio of exchangeable Ca to Na is 10 or less.

Vertic horizon—See definition of "v".

Lithic layer—This layer is consolidated bedrock (R) within the control section below a depth of 10 cm. The upper surface of a lithic layer is a lithic contact.

Mull—This form of zoogenous, forest humus consists of an intimate mixture of well-humified organic matter and mineral soil with crumb or granular structure that makes a gradual transition to the horizon underneath. Because of the activity of burrowing microfauna, (mostly earthworms), partly decomposed organic debris does not accumulate as a distinct layer (F layer) as in mor and moder. The organic matter content is usually 5–25% and the C:N ratio 12–18. It is a kind of Ah horizon.

Organic horizons

Organic horizons occur in Organic soils and commonly at the surface of mineral soils. They may occur at any depth beneath the surface in buried soils or overlying geologic deposits. They contain more than 17% organic C (about 30% or more organic matter) by weight. Two groups of these horizons are recognized, the O horizons (peat materials) and the L, F, and H horizons (folic materials).

O—This organic horizon is developed mainly from mosses, rushes, and woody materials. It is divided into the following subhorizons:

Of—This O horizon consists largely of fibric materials that are readily identifiable as to botanical origin. A fibric horizon (Of) has 40% or more of rubbed fiber by volume and a pyrophosphate index of 5 or more. If the rubbed fiber volume is 75% or more, the pyrophosphate criterion does not apply. Fiber is defined as the organic material retained on a 100-mesh sieve (0.15 mm), except for wood fragments that cannot be crushed in the hand and are larger than 2 cm in the smallest dimension. Rubbed fiber is the fiber that remains after rubbing a sample of the layer about 10 times between the thumb and forefinger. Fibric material usually is classified on the von Post scale of decomposition as class 1 to class 4. Three kinds of fibric horizons are named. Fennic horizons are derived from rushes, reeds, and sedges. Silvic horizons are derived from wood, moss with less than 75% of the volume being *Sphagnum* spp., and other herbaceous plants. Sphagnic horizons are derived from sphagnum mosses.

Om—This O horizon consists of mesic material, which is at a stage of decomposition intermediate between fibric and humic materials. The material is partly altered both physically and biochemically. It does not meet the requirements of either a fibric or a humic horizon, has a rubbed fiber content ranging from 10% to less than 40%, and has a pyrophosphate index of >3 and <5. Mesic material usually is classified on the von Post scale of decomposition as class 5 or 6.

Oh—This O horizon consists of humic material, which is at an advanced stage of decomposition. The horizon has the lowest amount of fiber, the highest bulk density, and the lowest saturated water-holding capacity of the O horizons. It is very stable and changes little physically or chemically with time unless it is drained. The rubbed fiber content is less than 10% by volume and the pyrophosphate index is 3 or less. Humic material usually is classified on the von Post scale of decomposition as class 7 or higher and rarely as class 6.

The methods of determining the properties of fibric, mesic, and humic materials are outlined later in this chapter.

Oco—This material is coprogenous earth, which is a limnic material that occurs in some Organic soils. It is deposited in water by aquatic organisms such as algae or derived from underwater and floating aquatic plants subsequently modified by aquatic animals.

L, F, and H—These organic horizons developed primarily from the accumulation of leaves, twigs, and woody materials with or without a minor component of mosses. They are normally associated with upland forested soils with imperfect drainage or drier.

L—This organic horizon is characterized by an accumulation of organic matter in which the original structures are easily discernible.

F—This organic horizon is characterized by an accumulation of partly decomposed organic matter. Some of the original structures are difficult to recognize. The material may be partly comminuted by soil fauna as in moder, or it may be a partly decomposed mat permeated by fungal hyphae as in mor.

H—This organic horizon is characterized by an accumulation of decomposed organic matter in which the original structures are indiscernible. This horizon differs from the F by having greater humification due chiefly to the action of organisms. It is frequently intermixed with mineral grains, especially near the junction with a mineral horizon.

Named layers and materials of Organic soils

Fibric, mesic, and humic materials were defined under Of, Om, and Oh. Some typical physical properties of fibric, mesic, and humic materials are listed below (Boelter 1969).

	Fibric material	Mesic material	Humic material
bulk density (Mg m^{-3})	<0.075	0.075–0.195	>0.195
total porosity (% vol)	>90	90–85	<85
0.01 M Pa H$_2$O content (% vol)	<48	48–70	>70
hydraulic conductivity (cm hr^{-1})	>6	6–0.1	<0.1

Limnic layer—This is a layer or layers, 5 cm or more thick, of coprogenous earth (sedimentary peat), diatomaceous earth, or marl. Except for some of the coprogenous earths containing more than 30% organic matter, most of these limnic materials are inorganic.

> Coprogenous earth is composed of aquatic plant debris modified by aquatic animals. It makes slightly viscous water suspensions and is slightly plastic but not sticky. The material shrinks upon drying to form clods that are difficult to rewet and commonly crack along horizontal planes. It has very few or no plant fragments recognizable to the naked eye, a pyrophosphate index of 5 or more, and a dry color value of less than 5. The cation exchange capacity (CEC) is less than 240 cmol kg^{-1}. It is designated Oco in horizon descriptions.

> Diatomaceous earth is composed mainly of the siliceous shells of diatoms. It has a matrix color value of 4±1, if not previously dried, that changes on drying to the permanent, light gray or whitish color of diatoms. The diatom shells can be identified by microscopic (440 ×) examination. Diatomaceous earth has a pyrophosphate index of 5 or more. It is frequently more nearly mineral than organic in composition. It is designated C in horizon descriptions.

> Marl is composed of the shells of aquatic animals and CaCO$_3$ precipitated in water. It has a moist color value of 6±1 and effervesces with dilute HCl. The color of the matrix usually does not change on drying. Marl contains too little organic matter to coat the carbonate particles. It is designated Ck in horizon descriptions.

Cumulic layer—This is a layer or layers of mineral material in Organic soils. Either the combined thickness of the mineral layers is more than 5 cm or a single mineral layer 5–30 cm thick occurs. One continuous mineral layer more than 30 cm thick in the middle or bottom tier is a terric layer.

Terric layer—This is an unconsolidated mineral substratum not underlain by organic matter, or one continuous unconsolidated mineral layer (with 17% or less organic C) more than 30 cm thick in the middle or bottom tiers underlain by organic matter, within a depth of 1.6 m from the surface.

Lithic layer—This is a consolidated mineral layer (bedrock) occurring within 10–160 cm of the surface of Organic soils.

Hydric layer—This is a layer of water that extends from a depth of not less than 40 cm from the organic surface to a depth of more than 1.6 m.

Tests for distinguishing organic layers

Unrubbed and rubbed fiber *See* methods 2.81 and 2.82 in *Manual on Soil Sampling and Methods of Analysis* (McKeague 1978).

Pyrophosphate index Place 1 g of sodium pyrophosphate in a small plastic, screw-topped container, add 4 ml of water and stir. With a syringe measure a 5 cm^3 sample of moist organic material as in method 2.81 and place it in the plastic container, stir, and let stand overnight. Mix the sample thoroughly the next day. Using tweezers insert one end of a strip of chromatographic paper about 5 cm long vertically into the suspension. With the screw top in place to avoid evaporation, let the paper strip stand in the suspension until it is wetted to the top. Remove the paper strip with tweezers, cut off and discard the soiled end, and blot the remainder of the strip on absorbent paper. Read the value and chroma of the strip using good illumination and viewing the strip through the holes in the Munsell chart. The pyrophosphate index is the difference between the Munsell value and chroma of the strip.

von Post scale of decomposition

In this field test squeeze a sample of the organic material within the closed hand. Observe the color of the solution that is expressed between the fingers, the nature of the fibers, and the proportion of the original sample that remains in the hand. Ten classes are defined as follows:

1—Undecomposed; plant structure unaltered; yields only clear water colored light yellow–brown.

2—Almost undecomposed; plant structure distinct; yields only clear water colored light yellow–brown.

3—Very weakly decomposed; plant structure distinct; yields distinctly turbid brown water, no peat substance passes between the fingers, residue not mushy.

4—Weakly decomposed; plant structure distinct; yields strongly turbid water, no peat substance escapes between the fingers, residue rather mushy.

5—Moderately decomposed; plant structure clear but becoming indistinct; yields much turbid brown water, some peat escapes between the fingers, residue very mushy.

6—Strongly decomposed; plant structure somewhat indistinct but clearer in the squeezed residue than in the undisturbed peat; about one-third of the peat escapes between the fingers, residue strongly mushy.

7—Strongly decomposed; plant structure indistinct but recognizable; about half the peat escapes between the fingers.

8—Very strongly decomposed; plant structure very indistinct; about two-thirds of the peat escapes between the fingers, residue almost entirely resistant remnants such as root fibers and wood.

9—Almost completely decomposed; plant structure almost unrecognizable; nearly all the peat escapes between the fingers.

10—Completely decomposed; plant structure unrecognizable; all the peat escapes between the fingers.

Rules concerning horizon and layer designations

1. Do not use the uppercase letters A, B, and O singly for horizons in pedon descriptions, but accompany them by a lowercase suffix (e.g., Ah, Bf, or Om) indicating the estimated nature of the modification of the horizon from the parent material. The horizon and layer designations L, F, H, R, and W may be used alone, and the horizon designation C may be used alone except when the material is affected by reducing conditions (Cg), cementation (Cc), salinity (Cs or Csa), $CaCO_3$, (Ck or Cca), or permafrost (Cz).

2. Unless otherwise specified, additional lowercase suffixes indicate a feature or features in addition to those characteristic of the defined main horizon. For example, the symbol Btg indicates that in addition to illuvial clay in the B horizon there is evidence of strong gleying. Some combinations such as Bmj are not used. In some cases, such as Bgf and Bhf, the combination of suffixes has a specific meaning that differs from the sum of the two suffixes used singly.

3. All horizons except A and B, and B and A may be vertically subdivided by consecutive numeral suffixes. The uppermost subdivision is indicated by the numeral 1; each successive subdivision with depth is indicated by the next numeral. This convention is followed regardless of whether or not the horizon subdivisions are interrupted by a horizon of a different character. For example, an acceptable subdivision of horizons would be Ae1, Bf, Ae2, Bt1, Bt2, C1, C2. In some instances it may be useful for sampling purposes to subdivide a single horizon, for example, Bm1-1, Bm1-2, Bm1-3.

4. Roman numerals are prefixed to the contrasting master horizon or layer designation (A, B, C) to indicate lithological discontinuities either within or below the solum. The first, or uppermost, material is not numbered, because the Roman numeral I is understood; the second contrasting material is designated II, and the others are numbered consecutively, with depth. Thus, for example, a sequence from the surface downward might be Ah, Bm, IIBm, IICa, IICk, IIICk.

Lithological discontinuity is due to a different mode of deposition, indicated by strongly contrasting textures (differing by two textural classes), or to a different mineralogical composition, indicating a difference in the material from which the horizons have formed. These contrasting materials have resulted form geologic deposition rather than pedogenic processes.

A change in the clay content associated with a Bt horizon (textural B) does not indicate a difference in parent material. The appearance of gravel, or a change in the ratio between the various sand separates, normally suggests a difference in parent materials. A different Roman numeral would not normally be needed for a buried soil, because the symbol (b) would be used. A stone line usually indicates the need for another Roman numeral. The material above the stone line is presumed to be transported. If transport was by wind or water, it is likely that during movement, material was sorted according to size.

All O horizons, which have developed from peat materials in a wetland environment, are considered to have resulted from only one mode of deposition. The same principle applies to L, F, and H horizons, which have developed from folic materials in a dominantly forest system. These horizons (O, L, F, and H) should not be designated as contrasting, even if they differ in botanical composition or degree of decomposition.

In some cases it is not necessary to use Roman numerals to show strongly contrasting horizons, for example if the horizon symbol already indicates the difference. Roman numerals are not required if the soil is composed of peat materials overlain by folic materials and underlain by mineral soil (L, F, Om, Oh, C) or if a mineral soil has a folic or peaty surface layer (L, F, Bm, BC, C; or Om, Ahg, Cg).

5. For transitional horizons uppercase letters are used as follows:
   If the transition is gradual, use AB, BC, etc.
   If the horizons are interfingered in the transitional zone, use A and B, B and C, etc.

The dominance of horizons in the transitional zone may be shown by order, AB or BA, etc. Lower case suffixes may also be added in some instances, e.g., ABg, ABgj, etc.

6. The designations for diagnostic horizons must be given in the sequence shown in the horizon definitions, e.g., Ahe not Aeh.

7. Where j is used, the suffix or suffixes that it modifies are written after other horizon suffixes, e.g., Btnj, Bntj. Bfjtj, Bfcjgj.

Although definitions are given for all horizon symbols, all possible combinations of horizon designations have not been covered and all horizons having the same designation do not have identical properties. Therefore horizon descriptions are necessary.

Need for precise definitions of horizons and layers

In many cases the definitions of soil horizons may seem almost pedantically specific. For example, the suffix "t" indicates a horizon enriched with silicate clay. However, a Bt horizon must have a clay content exceeding that of the overlying eluvial horizon by specified amounts depending upon texture. For example, if the clay content of the Ae is 10%, that of the Bt must be 13% or more; if the clay content of the Ae is 40%, that of the Bt must be 48% or more. Also a Bt horizon must have a thickness that meets specified limits and clay skins on ped surfaces or oriented clay in pores.

Some B horizons that are slightly enriched with silicate clay are not Bt horizons. For example, two pedons X and Y have clay contents as follows: X: Ae–20%, B–22%, C–21%; Y: Ae–20%, B–25%, C–21%. If there is no parent material discontinuity in either pedon and both have B horizons more than 5 cm thick with clay skins on ped surfaces, the B horizon of pedon Y is a Bt, but that of pedon X is not. The two pedons would probably be closely similar if they were derived from similar materials in the same area, but they would be classified in different orders (Luvisolic and Brunisolic) because one has a Bt horizon and the other does not. Yet the difference in the clay contents of the B horizons is only 3% and it could result from an analytical error. If the descriptions of the pedons indicated no difference in the development of B horizons, the particle size data would be checked. In most cases, clay skins would be thicker and more continuous in the B horizon of pedon Y than in that of pedon X.

From the point of view of the soil surveyor in the area, pedons X and Y are closely similar soils that belong in the same class even at the series level and certainly at the order level. However, for the soil taxonomist concerned with ordering the information on the whole population of soils in the country the classification of pedons X and Y in different orders is inevitable for two reasons. Soils have a continuum of properties, and specific limits are essential if soil taxonomy is to be applied in a uniform manner by users of the system. The classification of pedons X and Y in different orders does not imply that the use interpretations must be different nor that the pedons must be separated and delineated in mapping. This depends on the pattern of distribution of pedons X and Y and the scale of mapping. The indication that pedon X does not have a Bt horizon and that pedon Y does simply informs pedologists that the two B horizons have properties such that they are on opposite sides of the artificial line through the continuum of properties indicating the development of a horizon enriched in silicate clay. The alternatives of vague specifications of limits of diagnostic horizons or of relying on individual judgments lead to chaos in the ordering of soil information throughout the country.

Specific horizon definitions are based on a generalization of properties of soil horizons that are known to be representative of the main soil classes and reflect the kinds and degrees of soil development. Whenever possible, the specifications are based on observable or easily measurable properties. These horizon definitions are modified as the knowledge of soils increases and as concepts change. Because of the lack of sufficient knowledge, some soil horizons may not be defined adequately.

Chapter 3

Outline of the System and a Key to the Classification of a Pedon

Outline of the System

The Canadian system of soil classification at the order, great group, and subgroup levels is tabulated alphabetically according to the names of the orders. For each subgroup the abbreviation of the name is appended.

Order	Great Group	Subgroup
Brunisolic	Melanic Brunisol	Orthic Melanic Brunisol O.MB
		Eluviated Melanic Brunisol E.MB
		Gleyed Melanic Brunisol GL.MB
		Gleyed Eluviated Melanic Brunisol GLE.MB
	Eutric Brunisol	Orthic Eutric Brunisol O.EB
		Eluviated Eutric Brunisol E.EB
		Gleyed Eutric Brunisol GL.EB
		Gleyed Eluviated Eutric Brunisol GLE.EB
	Sombric Brunisol	Orthic Sombric Brunisol O.SB
		Eluviated Sombric Brunisol E.SB
		Duric Sombric Brunisol DU.SB
		Gleyed Sombric Brunisol GL.SB
		Gleyed Eluviated Sombric Brunisol GLE.SB
	Dystric Brunisol	Orthic Dystric Brunisol O.DYB
		Eluviated Dystric Brunisol E.DYB
		Duric Dystric Brunisol DU.DYB
		Gleyed Dystric Brunisol GL.DYB
		Gleyed Eluviated Dystric Brunisol GLE.DYB
Chernozemic	Brown Chernozem	Orthic Brown Chernozem O.BC
		Rego Brown Chernozem R.BC
		Calcareous Brown Chernozem CA.BC
		Eluviated Brown Chernozem E.BC
		Solonetzic Brown Chernozem SZ.BC
		Vertic Brown Chernozem V.BC
		Gleyed Brown Chernozem GL.BC
		Gleyed Rego Brown Chernozem GLR.BC
		Gleyed Calcareous Brown Chernozem GLCA.BC
		Gleyed Eluviated Brown Chernozem GLE.BC
		Gleyed Solonetzic Brown Chernozem GLSZ.BC
		Gleyed Vertic Brown Chernozem GLV.BC
	Dark Brown Chernozem	Orthic Dark Brown Chernozem O.DBC
		Rego Dark Brown Chernozem R.DBC
		Calcareous Dark Brown Chernozem CA.DBC
		Eluviated Dark Brown Chernozem E.DBC
		Solonetzic Dark Brown Chernozem SZ.DBC
		Vertic Dark Brown Chernozem V.DBC
		Gleyed Dark Brown Chernozem GL.DBC
		Gleyed Rego Dark Brown Chernozem GLR.DBC
		Gleyed Calcareous Dark Brown Chernozem GLCA.DBC
		Gleyed Eluviated Dark Brown Chernozem GLE.DBC
		Gleyed Solonetzic Dark Brown Chernozem GLSZ.DBC
		Gleyed Vertic Dark Brown Chernozem GLV.DBC
	Black Chernozem	Orthic Black Chernozem O.BLC
		Rego Black Chernozem R.BLC
		Calcareous Black Chernozem CA.BLC
		Eluviated Black Chernozem E.BLC
		Solonetzic Black Chernozem SZ.BLC

Order	Great Group	Subgroup
		Vertic Black Chernozem V.BLC
		Gleyed Black Chernozem GL.BLC
		Gleyed Rego Black Chernozem GLR.BLC
		Gleyed Calcareous Black Chernozem GLCA.BLC
		Gleyed Eluviated Black Chernozem GLE.BLC
		Gleyed Solonetzic Black Chernozem GLSZ.BLC
		Gleyed Vertic Black Chernozem GLV.BLC
	Dark Gray Chernozem	Orthic Dark Gray Chernozem O.DGC
		Rego Dark Gray Chernozem R.DGC
		Calcareous Dark Gray Chernozem CA.DGC
		Solonetzic Dark Gray Chernozem SZ.DGC
		Vertic Dark Gray Chernozem V.DGC
		Gleyed Dark Gray Chernozem GL.DGC
		Gleyed Rego Dark Gray Chernozem GLR.DGC
		Gleyed Calcareous Dark Gray Chernozem GLCA.DGC
		Gleyed Solonetzic Dark Gray Chernozem GLSZ.DGC
		Gleyed Vertic Dark Gray Chernozem GLV.DGC
Cryosolic	Turbic Cryosol	Orthic Eutric Turbic Cryosol OE.TC
		Orthic Dystric Turbic Cryosol OD.TC
		Brunisolic Eutric Turbic Cryosol BRE.TC
		Brunisolic Dystric Turbic Cryosol BRD.TC
		Gleysolic Turbic Cryosol GL.TC
		Regosolic Turbic Cryosol R.TC
		Histic Eutric Turbic Cryosol HE.TC
		Histic Dystric Turbic Cryosol HD.TC
		Histic Regosolic Turbic Cryosol HR.TC
	Static Cryosol	Orthic Eutric Static Cryosol OE.SC
		Orthic Dystric Static Cryosol OD.SC
		Brunisolic Eutric Static Cryosol BRE.SC
		Brunisolic Dystric Static Cryosol BRD.SC
		Luvisolic Static Cryosol L.SC
		Gleysolic Static Cryosol GL.SC
		Regosolic Static Cryosol R.SC
		Histic Eutric Static Cryosol HE.SC
		Histic Dystric Static Cryosol HD.SC
		Histic Regosolic Static Cryosol HR.SC
	Organic Cryosol	Fibric Organic Cryosol FI.OC
		Mesic Organic Cryosol ME.OC
		Humic Organic Cryosol HU.OC
		Terric Fibric Organic Cryosol TFI.OC
		Terric Mesic Organic Cryosol TME.OC
		Terric Humic Organic Cryosol THU.OC
		Glacic Organic Cryosol GC.OC
Gleysolic	Luvic Gleysol	Solonetzic Luvic Gleysol SZ.LG
		Fragic Luvic Gleysol FR.LG
		Humic Luvic Gleysol HU.LG
		Fera Luvic Gleysol FE.LG
		Orthic Luvic Gleysol O.LG
		Vertic Luvic Gleysol V.LG
	Humic Gleysol	Solonetzic Humic Gleysol SZ.HG
		Fera Humic Gleysol FE.HG
		Orthic Humic Gleysol O.HG
		Rego Humic Gleysol R.HG
		Vertic Humic Gleysol V.HG
	Gleysol	Solonetzic Gleysol SZ.G
		Fera Gleysol FE.G
		Orthic Gleysol O.G
		Rego Gleysol R.G
		Vertic Gleysol V.G
Luvisolic	Gray Brown Luvisol	Orthic Gray Brown Luvisol O.GBL
		Brunisolic Gray Brown Luvisol BR.GBL
		Podzolic Gray Brown Luvisol PZ.GBL
		Vertic Gray Brown Luvisol V.GBL
		Gleyed Gray Brown Luvisol GL.GBL

Order	Great Group	Subgroup
		Gleyed Brunisolic Gray Brown Luvisol GLBR.GBL
		Gleyed Podzolic Gray Brown Luvisol GLPZ.GBL
		Gleyed Vertic Gray Brown Luvisol GLV.GBL
	Gray Luvisol	Orthic Gray Luvisol O.GL
		Dark Gray Luvisol D.GL
		Brunisolic Gray Luvisol BR.GL
		Podzolic Gray Luvisol PZ.GL
		Solonetzic Gray Luvisol SZ.GL
		Fragic Gray Luvisol FR.GL
		Vertic Gray Luvisol V.GL
		Gleyed Gray Luvisol GL.GL
		Gleyed Dark Gray Luvisol GLD.GL
		Gleyed Brunisolic Gray Luvisol GLBR.GL
		Gleyed Podzolic Gray Luvisol GLPZ.GL
		Gleyed Solonetzic Gray Luvisol GLSZ.GL
		Gleyed Fragic Gray Luvisol GLFR.GL
		Gleyed Vertic Gray Luvisol GLV.GL
Organic	Fibrisol	Typic Fibrisol TY.F
		Mesic Fibrisol ME.F
		Humic Fibrisol HU.F
		Limnic Fibrisol LM.F
		Cumulic Fibrisol CU.F
		Terric Fibrisol T.F
		Terric Mesic Fibrisol TME.F
		Terric Humic Fibrisol THU.F
		Hydric Fibrisol HY.F
	Mesisol	Typic Mesisol TY.M
		Fibric Mesisol FI.M
		Humic Mesisol HU.M
		Limnic Mesisol LM.M
		Cumulic Mesisol CU.M
		Terric Mesisol T.M
		Terric Fibric Mesisol TFI.M
		Terric Humic Mesisol THU.M
		Hydric Mesisol HY.M
	Humisol	Typic Humisol TY.H
		Fibric Humisol FI.H
		Mesic Humisol ME.H
		Limnic Humisol LM.H
		Cumulic Humisol CU.H
		Terric Humisol T.H
		Terric Fibric Humisol TFI.H
		Terric Mesic Humisol TME.H
		Hydric Humisol HY.H
	Folisol	Hemic Folisol HE.FO
		Humic Folisol HU.FO
		Lignic Folisol LI.FO
		Histic Folisol HI.FO
Podzolic	Humic Podzol	Orthic Humic Podzol O.HP
		Ortstein Humic Podzol OT.HP
		Placic Humic Podzol P.HP
		Duric Humic Podzol DU.HP
		Fragic Humic Podzol FR.HP
	Ferro-Humic Podzol	Orthic Ferro-Humic Podzol O.FHP
		Ortstein Ferro-Humic Podzol OT.FHP
		Placic Ferro-Humic Podzol P.FHP
		Duric Ferro-Humic Podzol DU.FHP
		Fragic Ferro-Humic Podzol FR.FHP
		Luvisolic Ferro-Humic Podzol LU.FHP
		Sombric Ferro-Humic Podzol SM.FHP
		Gleyed Ferro-Humic Podzol GL.FHP
		Gleyed Ortstein Ferro-Humic Podzol GLOT.FHP
		Gleyed Sombric Ferro-Humic Podzol GLSM.FHP

Order	Great Group	Subgroup
	Humo-Ferric Podzol	Orthic Humo-Ferric Podzol O.HFP
		Ortstein Humo-Ferric Podzol OT.HFP
		Placic Humo-Ferric Podzol P.HFP
		Duric Humo-Ferric Podzol DU.HFP
		Fragic Humo-Ferric Podzol FR.HFP
		Luvisolic Humo-Ferric Podzol LU.HFP
		Sombric Humo-Ferric Podzol SM.HFP
		Gleyed Humo-Ferric Podzol GL.HFP
		Gleyed Ortstein Humo-Ferric Podzol GLOT.HFP
		Gleyed Sombric Humo-Ferric Podzol GLSM.HFP
Regosolic	Regosol	Orthic Regosol O.R
		Cumulic Regosol CU.R
		Gleyed Regosol GL.R
		Gleyed Cumulic Regosol GLCU.R
	Humic Regosol	Orthic Humic Regosol O.HR
		Cumulic Humic Regosol CU.HR
		Gleyed Humic Regosol GL.HR
		Gleyed Cumulic Humic Regosol GLCU.HR
Solonetzic	Solonetz	Brown Solonetz B.SZ
		Dark Brown Solonetz DB.SZ
		Black Solonetz BL.SZ
		Alkaline Solonetz A.SZ
		Gleyed Brown Solonetz GLB.SZ
		Gleyed Dark Brown Solonetz GLDB.SZ
		Gleyed Black Solonetz GLBL.SZ
	Solodized Solonetz	Brown Solodized Solonetz B.SS
		Dark Brown Solodized Solonetz DB.SS
		Black Solodized Solonetz BL.SS
		Dark Gray Solodized Solonetz DG.SS
		Gray Solodized Solonetz G.SS
		Gleyed Brown Solodized Solonetz GLB.SS
		Gleyed Dark Brown Solodized Solonetz GLDB.SS
		Gleyed Black Solodized Solonetz GLBL.SS
		Gleyed Dark Gray Solodized Solonetz GLDG.SS
		Gleyed Gray Solodized Solonetz GLG.SS
	Solod	Brown Solod B.SO
		Dark Brown Solod DB.SO
		Black Solod BL.SO
		Dark Gray Solod DG.SO
		Gray Solod G.SO
		Gleyed Brown Solod GLB.SO
		Gleyed Dark Brown Solod GLDB.SO
		Gleyed Black Solod GLBL.SO
		Gleyed Dark Gray Solod GLDG.SO
		Gleyed Gray Solod GLG.SO
	Vertic Solonetz	Brown Vertic Solonetz BV.SZ
		Dark Brown Vertic Solonetz DBV.SZ
		Black Vertic Solonetz BLV.SZ
		Gleyed Brown Vertic Solonetz GLBV.SZ
		Gleyed Dark Brown Vertic Solonetz GLDBV.SZ
		Gleyed Black Vertic Solonetz GLBLV.SZ
Vertisolic	Vertisol	Orthic Vertisol O.V
		Gleyed Vertisol GL.V
		Gleysolic Vertisol GLC.V
	Humic Vertisol	Orthic Humic Vertisol O.HV
		Gleyed Humic Vertisol GL.HV
		Gleysolic Humic Vertisol GLC.HV

Photographs of some subgroup profiles are shown in Figures 4–27.

Figure 4

Figure 5

Figure 6

Figure 7

Figure 4   Orthic Melanic Brunisol, Ontario.

Figure 5   Eluviated Eutric Brunisol, British Columbia.

Figure 6   Eluviated Dystric Brunisol, Saskatchewan.

Figure 7   Orthic Brown Chernozem, Alberta.

Figure 8

Figure 9

Figure 10

Figure 11

Figure 8   Orthic Black Chernozem, Alberta.

Figure 9   Orthic Eutric Turbic Cryosol, Northwest Territories.

Figure 10   Brunisolic Dystric Static Cryosol, Northwest Territories.

Figure 11   Glacic Organic Cryosol, Northwest Territories.

Figure 12

Figure 13

Figure 14

Figure 15

Figure 12 Orthic Humic Gleysol, Ontario.

Figure 13 Rego Gleysol, peaty phase, Ontario.

Figure 14 Fera Gleysol, Ontario.

Figure 15 Orthic Gray Brown Luvisol, Ontario.

Figure 16

Figure 17

Figure 18

Figure 19

Figure 16 Orthic Gray Luvisol, Alberta.

Figure 17 Mesic Fibrisol, Alberta.

Figure 18 Humic Mesisol, British Columbia.

Figure 19 Orthic Humic Podzol, Newfoundland.

Figure 20

Figure 21

Figure 22

Figure 23

Figure 20 Orthic Ferro-Humic Podzol, Quebec.

Figure 21 Orthic Humo-Ferric Podzol, Nova Scotia.

Figure 22 Orthic Regosol, Northwest Territories.

Figure 23 Cumulic Regosol, Northwest Teritories.

Figure 24

Figure 25

Figure 26

Figure 27

Figure 24 Brown Solodized Solonetz, Saskatchewan.

Figure 25 Brown Solod, Saskatchewan.

Figure 26 Orthic Humic Vertisol, Manitoba.

Figure 27 Orthic Vertisol, Saskatchewan.

Classifying a Pedon

The taxonomic class of a pedon can be determined by using the definitions and keys in this and other chapters. A knowledge of soil horizon definitions and soil terminology as presented in this publication is required to use the keys. Definitions of these terms in relation to soil climate are in Chapter 14.

Keys to soil orders, soil great groups, and soil subgroups are presented in this chapter. Complete definitions of each order, great group, and subgroup are contained in the chapter on each order. All the keys are arranged in a systematic order and are based on diagnostic soil criteria (or criterion). Keys for soil families and soil series within subgroups are not included in this publication. However, family differentiae are specified in Chapter 14, and an up-to-date list of approved series (CanSIS Canadian Soil Names File) is maintained by the Eastern Cereal and Oilseed Research Centre, Ottawa.

The following procedure should be used to key out the classification of a pedon:

1. Expose a vertical section through the pedon, describe the horizons, and take appropriate samples if the designation of some horizons must be checked by laboratory analysis. In some cases the final classification must await the laboratory data.

2. Study the *Key to Soil Orders* in this chapter and select the first order in the key that apparently includes the pedon involved.

3. Turn to the page indicated and verify that the definition of the soil order includes the pedon concerned. Then using the *Key to Soil Great Groups* (page 35), select the appropriate great group.

4. Turn to the page indicated and verify that the definition of the soil great group includes the pedon concerned. Then using the *Key to Soil Subgroups* (page 37), select the appropriate subgroup.

5. Turn to the page indicated and verify that the definition of the soil subgroup includes the pedon concerned.

6. To classify the pedon at the family level, turn to Chapter 14 and select the family designation that applies; e.g., loamy, mixed, shallow, acid, cold, humid family.

7. To classify the pedon at the series level, refer to a recent soil survey report for the area, or confer with the soil correlator of the province or territory.

For those familiar with Canadian soil taxonomy and the soils of the area under study, the detailed step-by-step procedure usually is not necessary. However, even experienced pedologists must await laboratory data before classifying some pedons. Periodically users of the system will encounter pedons that do not appear to fit any of the soil classes defined. If the pedon appears to represent a significant area (i.e., more than 800 ha) then describe it thoroughly, collect samples for analysis, indicate the properties that make it difficult to classify, and notify the soil correlator of the province or territory involved.

Key to Soil Orders

A. . Soils that have permafrost within 1 m of the surface or within 2 m if strongly cryoturbated. . . . . **Cryosolic order**, p. 73

B. . Other soils that either

　1. Have organic horizons (more than 17% organic C by weight) that extend from the surface to one of the following:

　　a. A depth of 60 cm or more if the surface layer is fibric material (Of) having a bulk density of <0.075 g cm^{-3}.

　　b. A depth of 40 cm or more if the surface layer consists of mesic or humic material (Om or Oh) having a bulk density ≥ 0.075 g cm^{-3}.

　　c. A depth of more than 40 cm if composed of folic materials (L, F, and H), or at least 10 cm if a lithic contact or fragmental materials are present. Folic materials must be more than twice the thickness of a mineral soil layer if the mineral layer is less than 20 cm thick.

or
2. Have one or more mineral horizons or layers within 40 cm of the surface in addition to the organic horizons (O) as follows:
    a. If a mineral horizon or layer thinner than 40 cm occurs at the surface, the underlying organic horizon or horizons must have a total thickness of at least 40 cm.
    b. If one or more mineral horizons or layers occur within 40 cm of the surface, the organic material must occupy more than 40 cm of the upper 80 cm of the control section . . . . **Organic order**, p. 97

C. . Other soils that have both a vertic horizon and a slickenside horizon, the top of which occurs within 1 m of the mineral surface . . . . **Vertisolic order**, p. 131

D. Other soils that have a podzolic B horizon and do not have a Bt horizon within 50 cm of the mineral surface . . . . **Podzolic order**, p. 107

E. . Other soils that are saturated with water and under reducing conditions either continuously or during some period of the year as indicated either by direct measurements of the water table and the oxidation–reduction status, or by any of the following morphological features within 50 cm of the mineral surface:
1. For all but red soil materials (hue 5YR or redder and color fades slowly on dithionite treatment).
    a. Chromas of 1 or less, without mottles, on ped surfaces or in the matrix if peds are lacking in materials that develop higher chromas under oxidizing conditions.
    b. Chromas of 2 or less, in hues of 10YR and 7.5YR, on ped surfaces or in the matrix if peds are lacking, accompanied by prominent mottles.
    c. Chromas of 3 or less, in hues yellower than 10YR, on ped surfaces or in the matrix if peds are lacking, accompanied by prominent mottles.
    d. Hues bluer than 10Y, with or without mottles, on ped surfaces or in the matrix if peds are lacking.
2. For red soil materials (hue 5YR or redder and color fades slowly on dithionite treatment).
    a. Distinct or prominent mottles at least 1 mm in diameter occupying at least 2% of the exposed, unsmeared 10 cm layer . . . . **Gleysolic order**, p. 81
F. . Other soils that have a solonetzic B horizon . . . . . . **Solonetzic order**, p. 121

G. Other soils that have a chernozemic A horizon and any one of the following:
1. No Ae horizon.
2. A weakly expressed Ae horizon (Aej) with a dry color value lower than 5.
3. An Ae horizon thinner than an overlying Ah or Ap horizon that does not appear to be eluviated.
4. An Ae horizon not more than 5 cm thick if the chernozemic A is eluviated (Ahe) as indicated by gray streaks and splotches when the soil is dry . . . . **Chernozemic order**, p. 61
H. Other soils that have a Bt horizon . . . . **Luvisolic order**, p. 89

I. . Other soils that have either Bm, Btj, or Bfj horizons at least 5 cm thick, or a Bf horizon less than 10 cm in thickness . . . . **Brunisolic order**, p. 53

J. . Other soils . . . . **Regosolic order**, p. 117

Key to Soil Great Groups

A. Cryosolic order

AA. Cryosolic soils that are formed primarily in organic materials and have permafrost within 1 m of the surface . . . .
. . . . **Organic Cryosol**, p. 79

AB. Other Cryosolic soils that are formed in mineral materials, have marked evidence of cryoturbation, and have permafrost within 2 m of the surface . . . . **Turbic Cryosol**, p. 74

AC. Other Cryosolic soils that are formed in mineral materials, do not have marked evidence of cryoturbation, and have permafrost within 1 m of the surface . . . . **Static Cryosol**, p. 77

B. Organic order

BA. Organic soils that are formed primarily in upland organic (folic) materials, generally of forest origin, and are rarely saturated with water . . . . **Folisol**, p. 104

BB. Other Organic soils that are formed in relatively undecomposed organic materials and have a dominantly fibric middle tier . . . . . .
. . . . **Fibrisol**, p. 99

BC. Other Organic soils that are formed in organic materials, are in an intermediate stage of decomposition, and have a dominantly mesic middle tier . . . . **Mesisol**, p. 101

BD. Other Organic soils that are formed in organic materials, are in an advanced stage of decomposition, and have a dominantly humic middle tier . . . . **Humisol**, p. 102

C. Vertisolic order

CA. Vertisolic soils that have either a surface color value of ≥3.5 dry if well to imperfectly drained or an Ah horizon <10 cm in thickness if poorly drained . . . . **Vertisol**, p. 132

CB. Other Vertisolic soils that have either a surface color value of <3.5 dry if well to imperfectly drained or an Ah horizon ≥10 cm in thickness if poorly drained . . . . **Humic Vertisol**, p. 133

D. Podzolic order

DA. Podzolic soils that have a Bh horizon ≥10 cm in thickness . . . .
. . . .**Humic Podzol**, p. 110

DB. Other Podzolic soils that have a Bhf horizon ≥10 cm in thickness . .
. . . . **Ferro-Humic Podzol**, p. 112

DC. Other Podzolic soils . . . . **Humo-Ferric Podzol**, p. 114

E. Gleysolic order

EA. Gleysolic soils that have a Btg horizon and usually have an eluvial horizon . . . . **Luvic Gleysol**, p. 84

EB. Other Gleysolic soils that have either an Ah horizon ≥10 cm in thickness or an Ap horizon ≥15 cm in thickness and have at least 2.0% organic C in the surface horizon . . . . **Humic Gleysol**, p. 85

EC. Other Gleysolic soils . . . . **Gleysol**, p. 86

F.   Solonetzic order

FA.   Solonetzic soils that have a slickenside horizon within 1 m of the mineral surface . . . . **Vertic Solonetz**, p. 128

FB.   Other Solonetzic soils that have an Ae horizon 2 cm in thickness with a distinct AB or BA horizon (disintegrating Bnt) . . . .
. . . . **Solod**, p. 126

FC.   Other Solonetzic soils that have an Ae horizon ≥2 cm in thickness . .
. . . . **Solodized Solonetz**, p. 124

FD.   Other Solonetzic soils . . . . **Solonetz**, p. 123

G.   Chernozemic order

GA.   Chernozemic soils that have a surface color value of 4.5–5.5 dry and a chroma usually >1.5 dry . . . . **Brown Chernozem**, p. 63

GB.   Other Chernozemic soils that have a surface color value of 3.5–4.5 dry and a chroma usually >1.5 dry . . . . **Dark Brown Chernozem**, p. 66

GC.   Other Chernozemic soils that have a surface color value of <3.5 dry and a chroma usually ≤1.5 dry . . . . **Black Chernozem**, p. 68

GD.   Other Chernozemic soils that have a surface color value of 3.5–4.5 (3.5–5.0 for Ap) dry, a chroma usually ≤1.5 dry, and characteristics indicating eluviation associated with soils formed under forest vegetation . . . . **Dark Gray Chernozem**, p. 69

H.   Luvisolic order

HA.   Luvisolic soils that have a forest mull Ah horizon and a mean annual soil temperature ≥8°C . . . . **Gray Brown Luvisol**, p. 91

HB.   Other Luvisolic soils . . . . **Gray Luvisol**, p. 93

I.   Brunisolic order

IA.   Brunisolic soils that have an Ah or Ap horizon ≥10 cm in thickness and pH of ≥5.5 (0.01 M CaCl₂) . . . . **Melanic Brunisol**, p. 55

IB.   Other Brunisolic soils that have either no Ah horizon or an Ah (or Ap) horizon <10cm in thickness and pH of ≥5.5 (0.01 M CaCl₂) .
. . . . **Eutric Brunisol**, p. 56

IC.   Brunisolic soils that have an Ah or Ap horizon ≥10 cm in thickness and pH of <5.5 (0.01 M CaCl₂) . . . . **Sombric Brunisol**, p. 57

ID.   Other Brunisolic soils that have no Ah horizon or an Ah (or Ap) <10 cm in thickness and pH of <5.5 (0.01 M CaCl₂) . . . .
. . . . **Dystric Brunisol**, p. 58

J.   Regosolic order

JA.   Regosolic soils that have an Ah or Ap horizon ≥10 cm in thickness. .
. . . . **Humic Regosol**, p. 119

JB.   Other Regosolic soils . . . . **Regosol**, p. 118

Key to Soil Subgroups

AA.   **Organic Cryosol**

AAA.   Organic Cryosols that have an ice layer >30 cm in thickness with the upper boundary within 1 m of the surface . . . .
. . . . **Glacic Organic Cryosol**, p. 79

AAB.   Other Organic Cryosols that have a mineral contact within 1 m of the surface and mainly fibric organic material above the contact . . . .
. . . . **Terric Fibric Organic Cryosol**, p. 79

AAC.   Other Organic Cryosols that have a mineral contact within 1 m of the surface and mainly mesic organic material above the contact . . . .
. . . . **Terric Mesic Organic Cryosol**, p. 79

AAD.   Other Organic Cryosols that have a mineral contact within 1 m of the surface and mainly humic organic material above the contact . . . .
. . . . **Terric Humic Organic Cryosol**, p. 79

AAE.   Other Organic Cryosols in which the organic material is dominantly fibric below a depth of 40 cm . . . . **Fibric Organic Cryosol**, p. 79

AAF.   Other Organic Cryosols in which the organic material is dominantly mesic below a depth of 40 cm . . . . **Mesic Organic Cryosol**, p. 79

AAG.   Other Organic Cryosols in which the organic material is dominantly humic below a depth of 40 cm . . . . **Humic Organic Cryosol**, p. 79

AB.   **Turbic Cryosol**

ABA.   Turbic Cryosols that have a gleyed layer similar to soils of the Gleysolic order . . . . **Gleysolic Turbic Cryosol**, p. 76

ABB.   Other Turbic Cryosols that have organic horizons in the upper 1 m of the solum, are >15 cm in thickness, and have a pH ≥5.5 in some or all of the B horizons . . . . **Histic Eutric Turbic Cryosol**, p. 76

ABC.   Other Turbic Cryosols that have organic horizons in the upper 1 m of the solum, are >15 cm in thickness, and have a pH <5.5 in some or all of the B horizons . . . . **Histic Dystric Turbic Cryosol**, p. 76

ABD.   Other Turbic Cryosols that have organic horizons in the upper 1 m of the solum, are >15 cm in thickness, and have no B horizons . . . .
. . . . **Histic Regosolic Turbic Cryosol**, p. 77

ABE.   Other Turbic Cryosols that have a Bm horizon ≥10 cm in thickness and a pH ≥5.5 in some or all of the B horizons . . . .
. . . . **Brunisolic Eutric Turbic Cryosol**, p. 75

ABF.   Other Turbic Cryosols that have a Bm horizon ≥10 cm in thickness and a pH <5.5 in some or all of the B horizons . . . .
. . . . **Brunisolic Dystric Turbic Cryosol**, p. 75

ABG.   Other Turbic Cryosols that have a Bm horizon <10 cm in thickness and a pH ≥5.5 in some or all of the B horizons . . . .
. . . . **Orthic Eutric Turbic Cryosol**, p. 74

ABH.   Other Turbic Cryosols that have a Bm horizon <10 cm in thickness and a pH <5.5 in some or all of the B horizons . . . .
. . . . **Orthic Dystric Turbic Cryosol**, p. 75

ABI.   Other Turbic Cryosols . . . . **Regosolic Turbic Cryosol**, p. 76

AC. **Static Cryosol**

ACA. Static Cryosols that have a gleyed layer similar to soils of the Gleysolic order . . . . **Gleysolic Static Cryosol**, p. 78

ACB. Other Static Cryosols that have organic horizons in the upper 1 m of the solum, are >15 cm in thickness, and have a pH ≥5.5 in some or all of the B horizons . . . .
. . . . **Histic Eutric Static Cryosol**, p. 78

ACC. Other Static Cryosols that have organic horizons in the upper 1 m of the solum, are >15 cm in thickness, and have a pH <5.5 in some or all of the B horizons . . . .
. . . . **Histic Dystric Static Cryosol**, p. 78

ACD. Other Static Cryosols that have organic horizons in the upper 1 m of the solum, are >15 cm in thickness, and have no B horizons . . . .
. . . . **Histic Regosolic Static Cryosol**, p. 78

ACE. Other Static Cryosols that have an eluvial horizon and a Bty horizon ≥10 cm in thickness . . . . **Luvisolic Static Cryosol**, p. 77

ACF. Other Static Cryosols that have a Bm horizon ≥10 cm in thickness and a pH ≥5.5 in some or all of the B horizons . . . .
. . . . **Brunisolic Eutric Static Cryosol**, p. 77

ACG. Other Static Cryosols that have a Bm horizon ≥10 cm in thickness and a pH <5.5 in some or all of the B horizons . . . .
. . . . **Brunisolic Dystric Static Cryosol**, p. 77

ACH. Other Static Cryosols that have a Bm horizon <10 cm in thickness and a pH ≥5.5 in some or all of the B horizons . . . .
. . . . **Orthic Eutric Static Cryosol**, p. 77

ACI. Other Static Cryosols that have a Bm horizon <10 cm in thickness and a pH <5.5 in some or all of the B horizons . . . .
. . . . **Orthic Dystric Static Cryosol**, p. 77

ACJ. Other Static Cryosols . . . . **Regosolic Static Cryosol**, p. 78

BA. **Folisol**

BAA. Folisols that have an O horizon >10 cm in thickness below the F or H horizons . . . . **Histic Folisol**, p. 105

BAB. Other Folisols that have F or H horizons composed primarily of woody materials . . . . **Lignic Folisol**, p. 105

BAC. Other Folisols that are composed primarily of moderately decomposed F horizon within the control section . . **Hemic Folisol**, p. 104

BAD. Other Folisols that are composed primarily of well-decomposed H horizon within the control section . . . . **Humic Folisol**, p. 104

BB. **Fibrisol**

BBA. Fibrisols that have a hydric layer . . . . **Hydric Fibrisol**, p. 101

BBB. Other Fibrisols that have a terric layer at least 30 cm in thickness beneath the surface tier and a humic layer >12 cm in thickness within the control section . . . . **Terric Humic Fibrisol**, p. 101

FAC. Other Vertic Solonetzs that have an Ah, Ahe, or Ap horizon with a color value of 3.5–4.5 dry, a chroma usually >1.5 dry, and faint to distinct mottles within 50 cm of the mineral soil surface . . . . **Gleyed Dark Brown Vertic Solonetz**, p. 129

FAD. Other Vertic Solonetzs that have an Ah, Ahe, or Ap horizon with a color value of 3.5–4.5 dry and a chroma usually >1.5 dry . . . . **Dark Brown Vertic Solonetz**, p. 128

FAE. Other Vertic Solonetzs that have an Ah, Ahe, or Ap horizon with a color value <3.5 dry, a chroma usually >1.5 dry, and faint to distinct mottles within 50 cm of the mineral soil surface . . . . **Gleyed Black Vertic Solonetz**, p. 129

FAF. Other Vertic Solonetzs that have an Ah, Ahe, or Ap horizon with a color value <3.5 dry and a chroma usually >1.5 dry . . . . **Black Vertic Solonetz**, p. 128

FB. Solod

FBA. Solods that have an Ah, Ahe, or Ap horizon with a color value >4.5 dry, a chroma usually >1.5, and faint to distinct mottles within 50 cm of the mineral soil surface . . . . **Gleyed Brown Solod**, p. 127

FBB. Other Solods that have an Ah, Ahe, or Ap horizon with a color value >4.5 dry and a chroma usually >1.5 . . . . **Brown Solod**, p. 126

FBC. Other Solods that have an Ah, Ahe, or Ap horizon with a color value of 3.5–4.5 dry, a chroma usually >1.5, and faint to distinct mottles within 50 cm of the mineral soil surface . . . . **Gleyed Dark Brown Solod**, p. 127

FBD. Other Solods that have an Ah, Ahe, or Ap horizon with a color value of 3.5–4.5 dry and a chroma usually >1.5 . . . . **Dark Brown Solod**, p. 126

FBE. Other Solods that have an Ah, Ahe, or Ap horizon with a color value <3.5 dry, a chroma usually >1.5, and faint to distinct mottles within 50 cm of the mineral soil surface . . . . **Gleyed Black Solod**, p. 127

FBF. Other Solods that have an Ah, Ahe, or Ap horizon with a color value <3.5 dry and a chroma usually >1.5 . . . . **Black Solod**, p. 127

FBG. Other Solods that have an Ah, Ahe, or Ap horizon with a color value of 3.5–4.5 dry, a chroma usually <2, and faint to distinct mottles within 50 cm of the mineral soil surface and are associated with a subhumid climate . . . . **Gleyed Dark Gray Solod**, p. 128

FBH. Other Solods that have an Ah, Ahe, or Ap horizon with a color value of 3.5–4.5 dry and a chroma usually <2 and are associated with a subhumid climate . . . . **Dark Gray Solod**, p. 127

FBI. Other Solods that have an Ah, Ahe, or Ap horizon with a color value >4.5 dry, a chroma usually <2, and faint to distinct mottles within 50 cm of the mineral soil surface and are associated with a subhumid climate . . . . **Gleyed Gray Solod**, p. 128

FBJ. Other Solods that have an Ah, Ahe, or Ap horizon with a color value >4.5 dry and a chroma usually <2 and are associated with a subhumid climate . . . . **Gray Solod**, p. 127

FC. Solodized Solonetz

FCA. Solodized Solonetzs that have an Ah, Ahe, or Ap horizon with a color value >4.5 dry, a chroma usually >1.5, and faint to distinct mottles within 50 cm of the mineral soil surface . . . . **Gleyed Brown Solodized Solonetz**, p. 126

FCB. Other Solodized Solonetzs that have an Ah, Ahe, or Ap horizon with a color value >4.5 dry and a chroma usually >1.5 . . . . **Brown Solodized Solonetz**, p. 125

FCC. Other Solodized Solonetzs that have an Ah, Ahe, or Ap horizon with a color value of 3.5–4.5 dry, a chroma usually >1.5, and faint to distinct mottles within 50 cm of the mineral soil surface . . . . **Gleyed Dark Brown Solodized Solonetz**, p. 126

FCD. Other Solodized Solonetzs that have an Ah, Ahe, or Ap horizon with a color value of 3.5–4.5 dry and a chroma usually >1.5 . . . . **Dark Brown Solodized Solonetz**, p. 125

FCE. Other Solodized Solonetzs that have an Ah, Ahe, or Ap horizon with a color value <3.5 dry, a chroma usually >1.5, and faint to distinct mottles within 50 cm of the mineral soil surface . . . . **Gleyed Black Solodized Solonetz**, p. 126

FCF. Other Solodized Solonetzs that have an Ah, Ahe, or Ap horizon with a color value <3.5 dry and a chroma usually >1.5 . . . . **Black Solodized Solonetz**, p. 125

FCG. Other Solodized Solonetzs that have an Ah, Ahe, or Ap horizon with a color value of 3.5–4.5 dry, a chroma usually <2, and faint to distinct mottles within 50 cm of the mineral soil surface and are associated with a subhumid climate . . . . **Gleyed Dark Gray Solodized Solonetz**, p. 126

FCH. Other Solodized Solonetzs that have an Ah, Ahe, or Ap horizon with a color value of 3.5–4.5 dry and a chroma usually <2 and are associated with a subhumid climate . . . . . . **Dark Gray Solodized Solonetz**, p. 125

FCI. Other Solodized Solonetzs that have an Ah, Ahe, or Ap horizon with a color value >4.5 dry, a chroma usually <2, and faint to distinct mottles within 50 cm of the mineral soil surface and are associated with a subhumid climate . . . . . **Gleyed Gray Solodized Solonetz**, p. 126

FCJ. Other Solodized Solonetzs that have an Ah, Ahe, or Ap horizon with a color value >4.5 dry and a chroma usually <2 and are associated with a subhumid climate . . . . **Gray Solodized Solonetz**, p. 125

FD. Solonetz

FDA. Solonetzs that have a strongly alkaline A horizon with pH (H$_2$O) >8.5 . . . . **Alkaline Solonetz**, p. 124

FDB. Other Solonetzs that have an Ah, Ahe, or Ap horizon with a color value >4.5 dry, a chroma usually >1.5, and faint to distinct mottles within 50 cm of the mineral soil surface . . . . **Gleyed Brown Solonetz**, p. 124

FDC. Other Solonetzs that have an Ah, Ahe, or Ap horizon with a color value >4.5 dry and a chroma usually >1.5 . . . . . **Brown Solonetz**, p. 123

FDD. Other Solonetzs that have an Ah, Ahe, or Ap horizon with a color value of 3.5–4.5 dry, a chroma usually >1.5, and faint to distinct mottles within 50 cm of the mineral soil surface . . . . **Gleyed Dark Brown Solonetz**, p. 124

FDE. Other Solonetzs that have an Ah, Ahe, or Ap horizon with a color value of 3.5–4.5 dry and a chroma usually >1.5 . . . . **Dark Brown Solonetz**, p. 124

FDF. Other Solonetzs that have an Ah, Ahe, or Ap horizon with a color value <3.5 dry, a chroma usually >1.5, and faint to distinct mottles within 50 cm of the mineral soil surface . . . . **Gleyed Black Solonetz**, p. 124

FDG. Other Solonetzs that have an Ah, Ahe, or Ap horizon with a color value <3.5 dry and a chroma usually >1.5 . . . . . . **Black Solonetz**, p. 124

GA. Brown Chernozem

GAA. Brown Chernozems that have a slickenside horizon within 1 m of the mineral soil surface and faint to distinct mottles within 50 cm of the mineral soil surface . . . . . **Gleyed Vertic Brown Chernozem**, p. 65

GAB. Other Brown Chernozems that have a slickenside horizon within 1 m of the mineral soil surface . . . . **Vertic Brown Chernozem**, p. 65

GAC. Other Brown Chernozems that have a Bnj, Bnjtj, or Btnj horizon and faint to distinct mottles within 50 cm of the mineral soil surface . . . . **Gleyed Solonetzic Brown Chernozem**, p. 65

GAD. Other Brown Chernozems that have a Bnj, Bnjtj, or Btnj horizon . . . . **Solonetzic Brown Chernozem**, p. 64

GAE. Other Brown Chernozems that have an eluvial horizon or horizons (Ahe, Ae, or Aej) at least 2 cm in thickness, usually underlain by a Btj or Bt horizon, and faint to distinct mottles within 50 cm of the mineral soil surface . . . . **Gleyed Eluviated Brown Chernozem**, p. 65

GAF. Other Brown Chernozems that have an eluvial horizon or horizons (Ahe, Ae, or Aej) at least 2 cm in thickness, usually underlain by a Btj or Bt horizon . . . . **Eluviated Brown Chernozem**, p. 64

GAG. Other Brown Chernozems that have a Bmk horizon at least 5 cm in thickness and faint to distinct mottles within 50 cm of the mineral soil surface . . . . **Gleyed Calcareous Brown Chernozem**, p. 65

GAH. Other Brown Chernozems that have a Bmk horizon at least 5 cm in thickness . . . . **Calcareous Brown Chernozem**, p. 64

GAI. Other Brown Chernozems that either lack a B horizon or have a B horizon <5 cm in thickness and have faint to distinct mottles within 50 cm of the mineral soil surface . . . . **Gleyed Rego Brown Chernozem**, p. 65

GAJ. Other Brown Chernozems that either lack a B horizon or have a B horizon <5 cm in thickness . . . . **Rego Brown Chernozem**, p. 64

GAK. Other Brown Chernozems that have faint to distinct mottles within 50 cm of the mineral soil surface . . . . . **Gleyed Brown Chernozem**, p. 65

GAL. Other Brown Chernozems . . . . **Orthic Brown Chernozem**, p. 63

GB. Dark Brown Chernozem

GBA. Dark Brown Chernozems that have a slickenside horizon within 1 m of the mineral soil surface and faint to distinct mottles within 50 cm of the mineral soil surface . . . . **Gleyed Vertic Dark Brown Chernozem**, p. 67

GC. Black Chernozem

GCF. Other Black Chernozems that have an eluvial horizon or horizons (Ahe, Ae, or Aej) at least 2 cm in thickness, usually underlain by a Btj or Bt horizon . . . . **Eluviated Black Chernozem**, p. 68

GCG. Other Black Chernozems that have a Bmk horizon at least 5 cm in thickness and faint to distinct mottles within 50 cm of the mineral soil surface . . . . **Gleyed Calcareous Black Chernozem**, p. 69

GCH. Other Black Chernozems that have a Bmk horizon at least 5 cm in thickness . . . . **Calcareous Black Chernozem**, p. 68

GCI. Other Black Chernozems that lack a B horizon or have a B horizon <5 cm in thickness and have faint to distinct mottles within 50 cm of the mineral soil surface . . . . **Gleyed Rego Black Chernozem**, p. 69

GCJ. Other Black Chernozems that lack a B horizon or have a B horizon <5 cm in thickness . . . . **Rego Black Chernozem**, p. 68

GCK. Other Black Chernozems that have faint to distinct mottles within 50 cm of the mineral soil surface . . . . . . **Gleyed Black Chernozem**, p. 69

GCL. Other Black Chernozems . . . . **Orthic Black Chernozem**, p. 68

GD. Dark Gray Chernozem

GDA. Dark Gray Chernozems that have a slickenside horizon within 1 m of the mineral soil surface and faint to distinct mottles within 50 cm of the mineral soil surface . . . . **Gleyed Vertic Dark Gray Chernozem**, p. 71

GDB. Other Dark Gray Chernozems that have a slickenside horizon within 1 m of the mineral soil surface . . . . . **Vertic Dark Gray Chernozem**, p. 70

GDC. Other Dark Gray Chernozems that have a Bnj, Bnjtj, or Btnj horizon and faint to distinct mottles within 50 cm of the mineral soil surface . . . . **Gleyed Solonetzic Dark Gray Chernozem**, p. 71

GDD. Other Dark Gray Chernozems that have a Bnj, Bnjtj, or Btnj horizon . . . . **Solonetzic Dark Gray Chernozem**, p. 70

GDE. Other Dark Gray Chernozems that have a Bmk horizon at least 5 cm in thickness and faint to distinct mottles within 50 cm of the mineral soil surface . . . . **Gleyed Calcareous Dark Gray Chernozem**, p. 71

GDF. Other Dark Gray Chernozems that have a Bmk horizon at least 5 cm in thickness . . . . **Calcareous Dark Gray Chernozem**, p. 70

GDG. Other Dark Gray Chernozems that either lack a B horizon or have a B horizon <5 cm in thickness and have faint to distinct mottles within 50 cm of the mineral soil surface . . . . **Gleyed Rego Dark Gray Chernozem**, p. 71

GDH. Other Dark Gray Chernozems that either lack a B horizon or have a B horizon <5 cm in thickness . . . . . . **Rego Dark Gray Chernozem**, p. 70

GDI. Other Dark Gray Chernozems that have faint to distinct mottles within 50 cm of the mineral soil surface . . . . **Gleyed Dark Gray Chernozem**, p. 71

GDJ. Other Dark Gray Chernozems . . . . . **Orthic Dark Gray Chernozem**, p. 70

HA.   Gray Brown Luvisol

HAA.   Gray Brown Luvisols that have a slickenside horizon within 1 m
        of the mineral soil surface and faint to distinct mottles within
        50 cm of the mineral soil surface . . . .
        . . . . **Gleyed Vertic Gray Brown Luvisol**, p. 93

HAB.   Other Gray Brown Luvisols that have a slickenside horizon within
        1 m of the mineral soil surface . . . . **Vertic Gray Brown Luvisol**, p. 92

HAC.   Other Gray Brown Luvisols that have a Podzolic B horizon ≥10 cm
        in thickness in the upper solum, a Bt horizon with its upper
        boundary within 50 cm from the mineral soil surface, and either
        distinct mottles within 50 cm of the mineral soil surface or prominent
        mottles at depths of 50–100 cm . . . .
        . . . . **Gleyed Podzolic Gray Brown Luvisol**, p. 93

HAD.   Other Gray Brown Luvisols that have a Podzolic B horizon ≥10 cm
        in thickness in the upper solum and a Bt horizon with its upper
        boundary within 50 cm of the mineral soil surface . . . .
        . . . . **Podzolic Gray Brown Luvisol**, p. 92

HAE.   Other Gray Brown Luvisols that have in the upper solum either a
        Bm horizon ≥5 cm in thickness with a chroma ≥3, or a Bf horizon
        <10 cm in thickness that does not extend below 15 cm of the mineral
        soil surface, and either distinct mottles within 50 cm of the mineral soil
        surface or prominent mottles at depths of 50–100 cm . . . .
        . . . . **Gleyed Brunisolic Gray Brown Luvisol**, p. 93

HAF.   Other Gray Brown Luvisols that have in the upper solum either a
        Bm horizon ≥5 cm in thickness with a chroma ≥3, or a Bf horizon
        <10 cm in thickness that does not extend below 15 cm of the mineral
        soil surface . . . . **Brunisolic Gray Brown Luvisol**, p. 92

HAG.   Other Gray Brown Luvisols that have either distinct mottles within
        50 cm of the mineral soil surface or prominent mottles at depths of
        50–100 cm . . . . **Gleyed Gray Brown Luvisol**, p. 92

HAH.   Other Gray Brown Luvisols . . . . **Orthic Gray Brown Luvisol**, p. 92

HB.   Gray Luvisol

HBA.   Gray Luvisols that have a slickenside horizon within 1 m of the
        mineral soil surface and faint to distinct mottles within 50 cm of the
        mineral soil surface . . . . **Gleyed Vertic Gray Luvisol**, p. 95

HBB.   Other Gray Luvisols that have a slickenside horizon within 1 m of
        the mineral soil surface . . . . **Vertic Gray Luvisol**, p. 94

HBC.   Other Gray Luvisols that have a fragipan either within or below the
        Bt horizon and have either distinct mottles within 50 cm of the
        mineral soil surface or prominent mottles at depths of 50–100 cm . . . . . .
        . . . . **Gleyed Fragic Gray Luvisol**, p. 95

HBD.   Other Gray Luvisols that have a fragipan either within or below the
        Bt horizon . . . . **Fragic Gray Luvisol**, p. 94

HBE.   Other Gray Luvisols that have a Podzolic B horizon ≥10 cm in
        thickness in the upper solum, a Bt horizon with its upper boundary
        within 50 cm of the mineral soil surface, and either distinct mottles
        within 50 cm of the mineral soil surface or prominent mottles at depths of
        50–100 cm . . . . **Gleyed Podzolic Gray Luvisol**, p. 95

48

HBF. Other Gray Luvisols that have a Podzolic B horizon ≥10 cm in thickness in the upper solum and a Bt horizon with its upper boundary within 50 cm of the mineral soil surface . . . . **Podzolic Gray Luvisol**, p. 94

HBG. Other Gray Luvisols that have an Ah or Ahe horizon ≥5 cm in thickness and either distinct mottles within 50 cm of the mineral soil surface or prominent mottles at depths of 50–100 cm . . . . **Gleyed Dark Gray Luvisol**, p. 95

HBH. Other Gray Luvisols that have either an Ah or Ahe horizon ≥5 cm in thickness . . . . **Dark Gray Luvisol**, p. 94

HBI. Other Gray Luvisols that have a Btnj horizon and either distinct mottles within 50 cm of the mineral soil surface or prominent mottles at depths of 50–100 cm . . . . **Gleyed Solonetzic Gray Luvisol**, p. 95

HBJ. Other Gray Luvisols that have a Btnj horizon . . . . **Solonetzic Gray Luvisol**, p. 94

HBK. Other Gray Luvisols that have in the upper solum either a Bm horizon ≥5 cm in thickness with a chroma ≥3, or a Bf horizon <10 cm in thickness that does not extend below 15 cm of the mineral soil surface, and either distinct mottles within 50 cm of the mineral soil surface or prominent mottles at depths of 50–100 cm . . . . **Gleyed Brunisolic Gray Luvisol**, p. 95

HBL. Other Gray Luvisols that have in the upper solum either a Bm horizon ≥5 cm in thickness with a chroma ≥3, or a Bf horizon <10 cm in thickness that does not extend below 15 cm of the mineral soil surface . . . . **Brunisolic Gray Luvisol**, p. 94

HBM. Other Gray Luvisols that have either distinct mottles within 50 cm of the mineral soil surface or prominent mottles at depths of 50–100 cm . . . . **Gleyed Gray Luvisol**, p. 95

HBN. Other Gray Luvisols . . . . **Orthic Gray Luvisol**, p. 93

IA. Melanic Brunisol

IAA Melanic Brunisols that have an Ae or Aej horizon ≥2 cm in thickness and either faint to distinct mottles within 50 cm of the mineral soil surface, or distinct to prominent mottles at depths of 50–100 cm . . . . **Gleyed Eluviated Melanic Brunisol**, p. 56

IAB. Other Melanic Brunisols that have an Ae or Aej horizon ≥2 cm in thickness . . . . **Eluviated Melanic Brunisol**, p. 56

IAC. Other Melanic Brunisols that have either faint to distinct mottles within 50 cm of the mineral soil surface, or distinct to prominent mottles at depths of 50–100 cm . . . . **Gleyed Melanic Brunisol**, p. 56

IAD. Other Melanic Brunisols . . . . **Orthic Melanic Brunisol**, p. 55

IB. Eutric Brunisol

IBA. Eutric Brunisols that have an Ae or Aej horizon ≥2 cm in thickness and either faint to distinct mottles within 50 cm of the mineral soil surface, or distinct to prominent mottles at depths of 50–100 cm . . . . **Gleyed Eluviated Eutric Brunisol**, p. 57

IBB. Other Eutric Brunisols that have an Ae or Aej horizon ≥2 cm in thickness . . . . **Eluviated Eutric Brunisol**, p. 57

JB.   **Regosol**

Chapter 4

Brunisolic Order

Great Group	Subgroup
Melanic Brunisol	Orthic Melanic Brunisol O.MB
	Eluviated Melanic Brunisol E.MB
	Gleyed Melanic Brunisol GL.MB
	Gleyed Eluviated Melanic Brunisol GLE.MB
Eutric Brunisol	Orthic Eutric Brunisol O.EB
	Eluviated Eutric Brunisol E.EB
	Gleyed Eutric Brunisol GL.EB
	Gleyed Eluviated Eutric Brunisol GLE.EB
Sombric Brunisol	Orthic Sombric Brunisol O.SB
	Eluviated Sombric Brunisol E.SB
	Duric Sombric Brunisol DU.SB
	Gleyed Sombric Brunisol GL.SB
	Gleyed Eluviated Sombric Brunisol GLE.SB
Dystric Brunisol	Orthic Dystric Brunisol O.DYB
	Eluviated Dystric Brunisol E.DYB
	Duric Dystric Brunisol DU.DYB
	Gleyed Dystric Brunisol GL.DYB
	Gleyed Eluviated Dystric Brunisol GLE.DYB

A diagrammatic representation of profiles of some subgroups of the Brunisolic order is given in Figure 28. Individual subgroups may include soils that have horizon sequences different from those shown. In the description of each subgroup, presented later in this chapter, a common horizon sequence is given; diagnostic horizons are underlined and some other commonly occurring horizons are listed.

Soils of the Brunisolic order have sufficient development to exclude them from the Regosolic order, but they lack the degree or kind of horizon development specified for soils of other orders. The central concept of the order is that of soils formed under forest and having brownish-colored Bm horizons, but the order also includes soils of various colors with both Ae horizons and B horizons having slight accumulations of either clay (Btj), or amorphous Al and Fe compounds (Bfj), or both. Soils having a Bf horizon less than 10 cm thick are a part of this order.

A Bm horizon may have any or all of the following: stronger chroma and redder hue than the underlying material, partial or complete removal of carbonates, slight illuviation based mainly on the occurrence of an overlying Ae horizon, a change in structure from that of the original material.

A Bm horizon may develop in materials of any color, such as gray, brown, black or red, and which vary in texture from gravel to clay.

Brunisolic soils include some that are calcareous to the surface and very slightly weathered, and others that are strongly acid and apparently weathered to about the same extent as the associated Podzolic soils. Most Brunisolic soils are well to imperfectly drained, but some that have been affected by seepage water are poorly drained although not strongly gleyed. They occur in a wide range of climatic and vegetative environments including Boreal Forest; mixed forest, shrubs, and grass; and heath and tundra.

Brunisolic soils have a Bm, Bfj, thin Bf, or Btj horizon at least 5 cm thick. They lack the diagnostic properties specified for soils of other orders. Thus they do not have any of the following: a solonetzic or podzolic B horizon, a Bt horizon, evidence of gleying as specified for soils of the Gleysolic order, organic horizons thicker than 40 cm if mesic or humic or 60 cm if fibric, permafrost within 1 m of the surface or 2 m if cryoturbated. Some Brunisolic soils have an Ah horizon, but

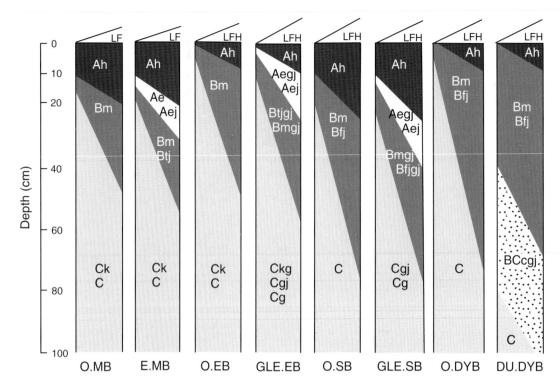

Figure 28 Diagrammatic horizon pattern of some subgroups of the Brunisolic order.

it does not meet the specifications for a chernozemic A either because of its inherent properties or the associated soil climate.

Distinguishing Brunisolic Soils from Soils of Other Orders

Guidelines for distinguishing Brunisolic soils from soils of other orders with which they might be confused follow. To a degree the Brunisolic order can be considered as an intergrade order between Regosolic soils and soils of several other orders. The distinctions are based more upon the degree than the kind of development.

Chernozemic Some Brunisolic soils and many Chernozemic soils have an Ah or dark-colored Ap horizon and either a Bm or a Btj horizon. These soils are classified as Chernozemic only if they have a chernozemic A horizon as defined in Chapter 2. For example, Melanic Brunisols of the St. Lawrence Lowlands have a chernozemic-like A horizon but are excluded from the Chernozemic order because their soil moisture regime is humid.

Soils of subalpine, alpine, northern boreal, and subarctic areas having Ah and Bm horizons are classified as Chernozemic if they have a chernozemic A horizon. Many similar soils in these areas do not have a chernozemic A either because of a soil moisture regime that is humid or wetter, a mean annual soil temperature that is colder than 0°C, or because of some inherent property of the Ah horizon such as low base saturation. These soils are classified as Brunisolic. Further studies of these soils may lead to improved criteria for differentiating Brunisolic and Chernozemic soils. Current information indicates that many Ah horizons of soils at high altitudes and latitudes have the following characteristics that differ from the Ah horizons of Chernozemic soils:

• Low degree of incorporation of organic matter with mineral material, moder. This includes the turfy A horizons of some alpine soils with bulk density less than 1.0 Mg m^{-3}.

In addition, some of these soils at high elevations have the following properties that differ from those of Chernozemic soils:

• Appreciable content of pyrophosphate-extractable Al and Fe in the Ah horizon, especially in soils containing volcanic ash.

• More than 1% organic C in the B horizon.

Luvisolic Luvisolic soils must have a Bt horizon, but Brunisolic soils do not. However, it is difficult to distinguish a Bt from a Btj

	Melanic Brunisol	Eutric Brunisol	Sombric Brunisol	Dystric Brunisol
Horizon	thick Ah or Ap (≥10 cm)	no (or thin) Ah or Ap (<10 cm)	thick Ah or Ap (≥10 cm)or	no (or thin) Ah or Ap (<10 cm)
pH	≥5.5	5.5	<5.5	<5.5
Ap color value	moist <4	moist ≥4	moist <4	moist ≥4

horizon and micromorphological examination may be required.

Podzolic Podzolic soils must have a podzolic B horizon and Brunisolic soils do not. However, the colors of some Bm and Bfj horizons are within the range of that of podzolic B horizons, and some of these horizons contain concentrations of amorphous complexes of Al and Fe with organic matter close to the minimum concentration diagnostic for Bf horizons. Therefore, chemical analysis is required to differentiate some Brunisolic soils from Podzolic soils. Soils having a Bf horizon thinner than 10 cm are classified as Brunisolic.

Regosolic Brunisolic soils must have a Bm, Bfj, thin Bf, or a Btj horizon at least 5 cm thick; Regosolic soils do not.

Cryosolic Cryosolic soils have permafrost within 1 m of the mineral surface or 2 m if strongly cryoturbated, but Brunisolic soils do not.

Vertisolic Vertisolic soils have both a vertic horizon (v) and a slickenside horizon (ss), but Brunisolic soils do not.

The Brunisolic order is divided into four great groups: Melanic Brunisol, Eutric Brunisol, Sombric Brunisol, and Dystric Brunisol based on soil reaction and the presence or absence of an Ah horizon, as indicated in the Brunisolic order chart.

Subgroups are separated on the basis of the kind and sequence of the horizons. Some former subgroup features (lithic, andic, turbic, and cryic) are now recognized taxonomically at either the family or series level, but different names are used for some of them. Alternatively, these features may be indicated as phases of subgroups, great groups, or orders.

---

[1] Diagnostic horizons are underlined.

Melanic Brunisol

These are Brunisolic soils having a dark-colored Ah horizon and a relatively high degree of base saturation as indicated by their pH. They occur typically under deciduous or mixed forest vegetation on materials of high base status in areas of boreal to mesic temperature class and humid moisture subclass, but they are not restricted to such environments. Many uncultivated Melanic Brunisols have a forest mull Ah horizon associated with the activity of soil fauna, especially earthworms.

Melanic Brunisols have an Ah horizon at least 10 cm in thickness or an Ap horizon 10 cm or more in thickness with a moist color value of less than 4, and either a Bm, Bfj, or Btj horizon at least 5 cm thick. The pH (0.01 M $CaCl_2$) is 5.5 or more in some part or all of the uppermost 25 cm of the B horizon, or in some part or all of the B horizon and the underlying material either to a total depth of 25 cm or to a lithic contact above that depth. Melanic Brunisols may have L, F, and H horizons and Ae or Aej horizons, but they do not have solonetzic or podzolic B horizons, or Bt horizons. The Ah horizons of some Melanic Brunisols have all of the properties diagnostic of a chernozemic A except the associated soil climate. If cultivated, these soils are classified as Melanic Brunisols if part of the Bm, Btj or Bfj horizon remains below the Ap and as Humic Regosols if the Ap includes all of the former B horizon.

Orthic Melanic Brunisol

Common horizon sequence: LF, Ah[1] , Bm , Ck or C

These soils have the general properties specified for the Brunisolic order and the Melanic Brunisol great group. Characteristically, they have a forest mull Ah horizon with fine to medium granular structure and a brownish-colored Bm horizon with a chroma of 3 or more. The color of the B horizon normally fades with depth. Commonly the C horizon is calcareous.

Orthic Melanic Brunisols are identified by the following properties:

1. They have either an Ah horizon 10 cm thick or more or an Ap horizon at least 10 cm thick with a moist color value less than 4. The A horizon does not meet the requirements of a chernozemic A.

2. They have a pH (0.01 M CaCl$_2$) of 5.5 or more as specified for the great group.

3. They have a Bm horizon at least 5 cm thick.

4. They lack an eluvial horizon, Ae or Aej, at least 2 cm thick.

5. They lack mottles that indicate gleying as specified for Gleyed Melanic Brunisols.

Orthic Melanic Brunisols and all other subgroups of Brunisolic soils may have a lithic contact within 50 cm of the surface or have turbic or andic features. These features are separated taxonomically at the family (lithic, some andic) or series (turbic) levels, or as phases of any taxonomic level above the family.

Eluviated Melanic Brunisol

Common horizon sequence: LF, Ah, Ae or Aej, Bm or Btj, Ck or C

These soils have the general properties specified for soils of the Brunisolic order and the Melanic Brunisol great group. They differ from Orthic Melanic Brunisols by having an eluvial horizon, Ae or Aej, at least 2 cm thick. The underlying horizon may be a Btj with thin clay skins on some surfaces or, less commonly, a Bfj. Otherwise, they have the diagnostic properties of Orthic Melanic Brunisols.

Gleyed Melanic Brunisol

Common horizon sequence: LF, Ah, Bmgj, Cgj or Cg or Ckg

These soils have the general properties specified for soils of the Brunisolic order and the Melanic Brunisol great group. They differ from Orthic Melanic Brunisols by having faint to distinct mottles within 50 cm of the mineral surface, or distinct or prominent mottles at depths of 50–100 cm. Otherwise, they have

the diagnostic properties of Orthic Melanic Brunisols.

Gleyed Eluviated Melanic Brunisol

Common horizon sequence: LF, Ah, Ae or Aej, Bmgj or Btjgj, Cgj or Cg or Ckg

These soils have the general properties specified for soils of the Brunisolic order and the Melanic Brunisol great group. They differ from Eluviated Melanic Brunisols by having mottles that indicate gleying. Gleyed Eluviated Melanic Brunisols have either an Ae or an Aej horizon at least 2 cm thick, and mottles as specified for Gleyed Melanic Brunisols.

Eutric Brunisol

These are Brunisolic soils that have a relatively high degree of base saturation as indicated by their pH and lack a well-developed mineral–organic surface horizon. They occur mainly on parent material of high base status under forest or shrub vegetation in a wide range of climates.

Eutric Brunisols have either a Bm, Bfj, or Btj horizon at least 5 cm thick, and a pH (0.01 M CaCl$_2$) of 5.5 or more in some part or all of the uppermost 25 cm of the B horizon, or some part or all of the B horizon and the underlying material either to a total depth of 25 cm or to a lithic contact above that depth. Eutric Brunisols may have L, F, and H horizons, Ae or Aej horizons, and an Ah horizon less than 10 cm thick, but they do not have either Bt or podzolic B horizons. If cultivated, these soils are classified as Melanic Brunisols if the Ap horizon is at least 10 cm thick and has a moist color value less than 4 and part of the Bm, Bfj, or Btj remains below the Ap. They are classed as Eutric Brunisols if the Ap horizon does not meet the above specifications and part of the Bm horizon remains below the Ap. If the Ap includes all of the former B horizon, they are classified as Humic Regosols or Regosols, depending on the Ap.

Orthic Eutric Brunisol

Common horizon sequence: LFH, Bm, Ck or C

These soils have the general properties specified for the Brunisolic order and the Eutric Brunisol great group. Usually they have one or more organic surface horizons overlying a brownish-colored, base-saturated B horizon. The C horizon is commonly calcareous.

Orthic Eutric Brunisols are identified by the following properties:

1. They have a pH (0.01 M CaCl$_2$) of 5.5 or more as specified for the great group.

2. They have a Bm horizon at least 5 cm thick.

3. They lack an eluvial horizon, Ae or Aej, at least 2 cm thick.

4. They lack mottles that indicate gleying as specified for Gleyed Melanic Brunisols.

5. They lack an Ah horizon at least 10 cm thick and an Ap horizon at least 10 cm thick with a moist color value of 4 or less.

Eluviated Eutric Brunisol

Common horizon sequence: LFH, <u>Ae</u> or <u>Aej</u>, <u>Bm</u> or <u>Btj</u>, Ck or C

These soils have the general properties specified for soils of the Brunisolic order and the Eutric Brunisol great group. They differ from Orthic Eutric Brunisols by having an eluvial horizon, Ae or Aej, at least 2 cm thick. The underlying horizon may be a Btj with thin clay skins on some surfaces or, less commonly, a Bfj. Otherwise, they have the diagnostic properties of Orthic Eutric Brunisols.

Gleyed Eutric Brunisol

Common horizon sequence: LFH, <u>Bmgj</u>, Cgj or Cg or Ckg

These soils have the general properties specified for soils of the Brunisolic order and the Eutric Brunisol great group. They differ from Orthic Eutric Brunisols by having faint to distinct mottles within 50 cm of the mineral surface, or distinct or prominent mottles at depths of 50–100 cm. Otherwise, they have the diagnostic properties of Orthic Eutric Brunisols.

Gleyed Eluviated Eutric Brunisol

Common horizon sequence: LFH, <u>Ae</u> or <u>Aegj</u>, <u>Bmgj</u> or <u>Btjgj</u>, Cgj or Cg or Ckg

These soils have the general properties specified for soils of the Brunisolic order and the Eutric Brunisol great group. They differ from Eluviated Eutric Brunisols by having mottles that indicate gleying. Gleyed

Eluviated Eutric Brunisols have either an Ae or an Aej horizon at least 2 cm thick and mottles as specified for Gleyed Eutric Brunisols.

Sombric Brunisol

These are acid Brunisolic soils having a dark-colored Ah horizon and a relatively low base saturation as indicated by their pH. Minor areas of soils of this great group occur in association with the more widely spread Podzolic soils.

Sombric Brunisols have an Ah 10 cm or more in thickness or an Ap horizon 10 cm or more in thickness with a moist color value of less than 4, and a Bm, Bfj, thin Bf, or Btj horizon at least 5 cm thick. The pH (0.01 M CaCl$_2$) is less than 5.5 throughout the uppermost 25 cm of the B horizon, or throughout the B horizon and the underlying material to a total depth of at least 25 cm or to a lithic contact above that depth. Sombric Brunisols may have L, F, and H horizons and Ae or Aej horizons, but they do not have solonetzic or podzolic B horizons or Bt horizons.

Orthic Sombric Brunisol

Common horizon sequence: LFH, <u>Ah</u>, <u>Bm</u>, C

These soils have the general properties specified for the Brunisolic order and the Sombric Brunisol great group. Usually they have an organic layer at the surface, a dark grayish brown to black Ah horizon, a brown acid B horizon, and an acid C horizon.

Orthic Sombric Brunisols are identified by the following properties:

1. They have an Ah horizon at least 10 cm thick or an Ap horizon at least 10 cm thick with a moist color value less than 4.

2. They have a pH (0.01 M CaCl$_2$) of less than 5.5 as specified for the great group.

3. They have a Bm horizon at least 5 cm thick.

4. They lack an eluvial horizon, Ae or Aej, at least 2 cm thick.

5. They lack mottles that indicate gleying as specified for Gleyed Sombric Brunisols.

6. They lack a duric horizon.

Eluviated Sombric Brunisol

Common horizon sequence: LFH, <u>Ah</u>, <u>Ae</u> or <u>Aej</u>, <u>Bm</u> or <u>Bfj</u>, C

These soils have the general properties specified for soils of the Brunisolic order and the Sombric Brunisol great group. They differ from Orthic Sombric Brunisols by having an eluvial horizon, Ae or Aej, at least 2 cm thick. The underlying horizon may be a Btj with thin clay skins on some surfaces or, more commonly, a Bfj. Otherwise, these soils have the diagnostic properties of Orthic Sombric Brunisols.

Duric Sombric Brunisol

Common horizon sequence: LFH, <u>Ah</u>, <u>Bm</u> or <u>Bfj</u>, <u>Bc</u> or <u>BCc</u>, C

These soils have the general properties specified for the Brunisolic order and the Sombric Brunisol great group. They differ from Orthic Sombric Brunisols by having a duric horizon within the control section. They may also have Ae and Btj or Bfj horizons and mottles that indicate gleying. A duric horizon is a strongly cemented acidic horizon that does not meet the requirements of a podzolic B and usually has a color similar to that of the parent material. This horizon has either an abrupt or a clear upper boundary and a diffuse lower boundary usually at least 50 cm below. Air-dry clods do not slake when immersed in water.

Gleyed Sombric Brunisol

Common horizon sequence: LFH, <u>Ah</u>, <u>Bmgj</u> or <u>Bfjgj</u>, Cgj or Cg

These soils have the general properties specified for soils of the Brunisolic order and the Sombric Brunisol great group. They differ from Orthic Sombric Brunisols by having faint to distinct mottles within 50 cm of the mineral surface or distinct or prominent mottles at depths of 50–100 cm. Otherwise, these soils have the diagnostic properties of Orthic Sombric Brunisols.

Gleyed Eluviated Sombric Brunisol

Common horizon sequence: LFH, <u>Ah</u>, <u>Aej</u> or <u>Aegj</u>, <u>Bmgj</u> or <u>Bfjgj</u>, Cgj or Cg

These soils have the general properties specified for soils of the Brunisolic order and the Sombric Brunisol great group. They differ from Eluviated Sombric Brunisols by having mottles that indicate gleying. Gleyed Eluviated Sombric Brunisols have either an Ae or an Aej horizon at least 2 cm thick and

mottles as specified for Gleyed Sombric Brunisols.

Dystric Brunisol

These are acid Brunisolic soils that lack a well-developed mineral–organic surface horizon. They occur widely, usually on parent materials of low base status and typically under forest vegetation.

Dystric Brunisols have a Bm, Bfj, thin Bf, or Btj horizon at least 5 cm thick, and a pH (0.01 M $CaCl_2$) of less than 5.5 throughout the upper 25 cm of the B horizon, or throughout the B horizon and the underlying material to a total depth of at least 25 cm or to a lithic contact above this depth. Dystric Brunisols may have L, F, and H horizons; an Ae or Aej horizon, and an Ah horizon less than 10 cm thick, but they do not have either a Bt or a podzolic B horizon. If cultivated, these soils are classified as Sombric Brunisols when the Ap horizon is at least 10 cm thick and has a moist color value less than 4 and part of the B horizon remains below the Ap. If the Ap horizon does not meet the above specifications but part of the B horizon remains below the Ap, then the soils are Dystric Brunisols. If the Ap includes all of the former B horizon, the soils are classified as Humic Regosol or Regosol, depending on the Ap horizon.

Orthic Dystric Brunisol

Common horizon sequence: LFH, <u>Bm</u>, C

These soils have the general properties specified for the Brunisolic order and the Dystric Brunisol great group. Usually they have organic surface horizons and brownish-colored, acid B horizons overlying acid C horizons.

Orthic Dystric Brunisols are identified by the following properties:

1. They have a pH (0.01 M $CaCl_2$) of less than 5.5 as specified for the great group.

2. They have a Bm horizon at least 5 cm thick.

3. They lack an eluvial horizon, Ae or Aej, at least 2 cm thick.

4. They lack mottles that indicate gleying as specified for Gleyed Dystric Brunisols.

5. They lack a duric horizon.

6. They lack an Ah horizon at least 10 cm thick and an Ap horizon at least 10 cm thick with a moist color value of 4 or less.

Eluviated Dystric Brunisol

Common horizon sequence: LFH, <u>Ae</u> or <u>Aej</u>, <u>Bm</u> or <u>Bfj</u>, C

These soils have the general properties specified for soils of the Brunisolic order and the Dystric Brunisol great group. They differ from Orthic Dystric Brunisols by having an eluvial horizon, Ae or Aej, at least 2 cm thick. The underlying horizon is commonly a Bfj, but it may be a Btj or a Bm. Otherwise, they have the diagnostic properties of Orthic Dystric Brunisols.

Duric Dystric Brunisol

Common horizon sequence: LFH, <u>Bm</u> or <u>Bfj</u>, <u>Bc</u> or <u>BCc</u>, C

These soils have the general properties specified for soils of the Brunisolic order and the Dystric Brunisol great group. They differ from Orthic Dystric Brunisols by having a duric horizon within the control section. Also they may have Ae and Btj or Bfj horizons and mottles that indicate gleying.

Gleyed Dystric Brunisol

Common horizon sequence: LFH, <u>Bmgj</u> or <u>Bfjgj</u>, Cgj or Cg

These soils have the general properties specified for soils of the Brunisolic order and the Dystric Brunisol great group. They differ from Orthic Dystric Brunisols by having faint to distinct mottles within 50 cm of the mineral surface, or distinct or prominent mottles at depths of 50–100 cm. Otherwise, they have the diagnostic properties of Orthic Dystric Brunisols.

Gleyed Eluviated Dystric Brunisol

Common horizon sequence: LFH, <u>Ae</u> or <u>Aej</u>, <u>Bmgj</u> or <u>Bfjgj</u>, Cgj or Cg

These soils have the general properties specified for soils of the Brunisolic order and the Dystric Brunisol great group. They differ from Eluviated Dystric Brunisols by having mottles that indicate gleying. Gleyed Eluviated Dystric Brunisols have either an Ae or an Aej horizon at least 2 cm thick and mottles as specified for Gleyed Dystric Brunisols.

Chapter 5

Chernozemic Order

Great Group	Subgroup
Brown Chernozem	Orthic Brown Chernozem O.BC
	Rego Brown Chernozem R.BC
	Calcareous Brown Chernozem CA.BC
	Eluviated Brown Chernozem E.BC
	Solonetzic Brown Chernozem SZ.BC
	Vertic Brown Chernozem V.BC
	Gleyed Brown Chernozem GL.BC
	Gleyed Rego Brown Chernozem GLR.BC
	Gleyed Calcareous Brown Chernozem GLCA.BC
	Gleyed Eluviated Brown Chernozem GLE.BC
	Gleyed Solonetzic Brown Chernozem GLSZ.BC
	Gleyed Vertic Brown Chernozem GLV.BC
Dark Brown Chernozem	Subgroups the same as for Brown Chernozem except for the great group name.
Black Chernozem	Subgroups the same as for Brown Chernozem except for the great group name.
Dark Gray Chernozem	Orthic Dark Gray Chernozem O.DGC
	Rego Dark Gray Chernozem R.DGC
	Calcareous Dark Gray Chernozem CA.DGC
	Solonetzic Dark Gray Chernozem SZ.DGC
	Vertic Dark Gray Chernozem V.DGC
	Gleyed Dark Gray Chernozem GL.DGC
	Gleyed Rego Dark Gray Chernozem GLR.DGC
	Gleyed Calcareous Dark Gray Chernozem GLCA.DGC
	Gleyed Solonetzic Dark Gray Chernozem GLSZ.DGC
	Gleyed Vertic Dark Gray Chernozem GLV.DGC

A diagrammatic representation of profiles of some subgroups of the Chernozemic order is given in Figure 29. Individual subgroups may include soils that have horizon sequences different from those shown. In the description of each subgroup, presented later in this chapter, a common horizon sequence is given; diagnostic horizons are underlined and some other commonly occurring horizons are listed.

The general concept of the Chernozemic order is that of well to imperfectly drained soils having surface horizons darkened by the accumulation of organic matter from the decomposition of xerophytic or mesophytic grasses and forbs representative of grassland communities or of grassland–forest communities with associated shrubs and forbs. The major area of Chernozemic soils is the cool, subarid to subhumid Interior Plains of Western Canada. Minor areas of Chernozemic soils occur in some valleys and mountain slopes in the Cordilleran Region extending in some cases beyond the tree line. Most Chernozemic soils are frozen during some period each winter and their sola are dry at some period each summer. Their mean annual soil temperature is $\geq 0°C$ but usually less than 5.5°C. However, some Chernozemic soils in dry valleys of British Columbia have higher temperatures.

The specific definition is as follows: soils of the Chernozemic order have an A horizon in which organic matter has accumulated (Ah, Ahe, Ap) and that meets the requirements of a chernozemic A horizon. A chernozemic A horizon has the following properties:

1. It is at least 10 cm thick or is thick and dark enough to provide 10 cm of surface material that meets the color criteria given in 2 and 3.

2. It has a color value darker than 5.5 dry and 3.5 moist and has a chroma of less than 3.5 moist.

61

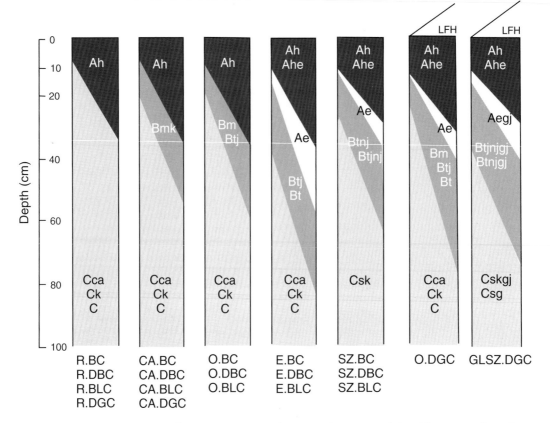

Figure 29 Diagrammatic horizon pattern of some subgroups of the Chernozemic order.

3. It has a color value at least one Munsell unit darker than that of the IC horizon.

4. It contains 1–17% organic C and its C:N ratio is less than 17.

5. Characteristically it has neither massive structure and hard consistence nor single-grained structure, when dry.

6. It has a base saturation (neutral salt) of more than 80% and Ca is the dominant exchangeable cation.

7. It is restricted to soils having a mean annual soil temperature of 0°C or higher and a soil moisture regime subclass drier than humid. Chernozemic soils may have an Ae horizon and a Bm or a Bt horizon.

They do not have any of the following: solonetzic B, podzolic B, evidence of gleying strongly enough expressed to meet the criteria of Gleysolic soils, or permafrost within 2 m of the surface.

Distinguishing Chernozemic Soils from Soils of Other Orders

Throughout the major area of Chernozemic soils in Canada there is little difficulty in distinguishing them from soils of other orders. However, soils of several other orders may have dark-colored Ah horizons. The bases for distinguishing Chernozemic soils from such soils are outlined below.

Solonetzic These soils have a solonetzic B horizon, but Chernozemic soils do not.

Luvisolic Some Dark Gray Luvisolic and some Chernozemic soils have all of the following: a chernozemic A horizon, an Ae horizon, a Bt horizon, and a subhumid soil moisture regime. The classification of these soils at the order level is done according to these guidelines:

1. If the chernozemic A horizon is eluviated as evidenced by gray streaks and splotches when the soil is dry, and if the Ae extends to a depth of at least 5 cm below the overlying Ah, Ahe, or Ap horizon, the soil is Luvisolic (Dark Gray Luvisol).

2. If the chernozemic A is not eluviated as described above, the soil is classified as Chernozemic unless the Ae horizon has a dry color value higher than 5 and a thickness greater than that of the Ah. In the latter case, the soil is classified as Luvisolic (Dark Gray Luvisol).

Podzolic These soils have a podzolic B horizon, but Chernozemic soils do not.

Brunisolic Brunisolic soils having dark-colored, mineral–organic surface horizons do not have a chernozemic A horizon either because their mean annual soil temperature is below 0°C, or because their soil moisture regime is humid or wetter, or the A horizon is acid or has a high C:N ratio. However, a degree of ambiguity remains with respect to the separation of some Melanic Brunisols from Chernozemic soils, especially in subalpine and alpine areas. Further work on these soils may disclose definitive differences between their A horizons and those of Chernozemic soils (see Chapter 4, Brunisolic order).

Regosolic These soils do not have a chernozemic A horizon.

Gleysolic Some of these soils have a chernozemic A horizon. They are excluded from the Chernozemic order because of evidence of reduction and poor drainage as specified in the Gleysolic order definition.

Vertisolic Some Chernozemic soils have a slickenside horizon but Vertisolic soils have both a vertic horizon and a slickenside horizon. Chernozemic soils do not.

Chernozemic soils are divided into four great groups: Brown Chernozem, Dark Brown Chernozem, Black Chernozem, and Dark Gray Chernozem mainly based on the color of the chernozemic A horizon, which reflects differences in the nature and amount of organic matter incorporated with the mineral material because of differences in climate and vegetation. The bases for great group separations are tabulated in the Chernozemic order chart.

Subgroups are separated based on the kind and sequence of the horizons indicating conformity with the central concept of the great group or intergrades to soils of other orders, or additional features. Some of the former subgroup features (carbonated, saline, and lithic) are now recognized taxonomically at either the family or series level. They may be indicated also as phases of subgroups, great groups, or orders. Features formerly referred to as Grumic are now recognized at either the order level (Vertisolic soils) or at the subgroup level (Vertic intergrades). The former Solonetzic and Solodic subgroups are now combined into a single subgroup Solonetzic, which includes all the intergrades to the Solonetzic order.

Brown Chernozem

These are soils that occur in the most arid segment of the climatic range of Chernozemic soils and have brownish-colored A horizons. They are associated with xerophytic and mesophytic grass and forb vegetation. In virgin Brown Chernozems the upper part of the Ah horizon is commonly as light as or lighter in color value than the lower part of the Ah or the upper B horizon.

Brown Chernozems have chernozemic Ah or Ap horizons with color values darker than 3.5 moist and 4.5–5.5 dry. The chroma of the A horizon is usually higher than 1.5. The soil climate of this great group is typically cold, rarely mild, and is subarid to semiarid (*see* Chapter 14 on soil family).

Orthic Brown Chernozem

Common horizon sequence: Ah, Bm, Cca or Ck

The Orthic Brown Chernozem subgroup may be thought of as the central concept of Brown Chernozems. It encompasses the properties specified for the Chernozemic order and the Brown Chernozem great group. Orthic Brown Chernozems are well drained and their B horizons are usually brown with prismatic macrostructure. The prismatic peds are

Chernozemic Order

	Brown Chernozem	Dark Brown Chernozem	Black Chernozem	Dark Gray Chernozem
Ah or Ap horizon	present	present	present	present
Color value (dry)	4.5–5.5	3.5–4.5	<3.5	3.5–4.5 (Ap 3.5–5)
Chroma (dry)	usually >1.5	usually >1.5	usually ≤1.5	usually ≤1.5
Climate	subarid to semiarid	semiarid	subhumid	subhumid

usually coarse in coarse-textured soils. With increasing clay content, the peds become smaller and have an increasing tendency to crush to fine blocky and granular aggregates. Thin clay coatings occur occasionally to commonly on ped surfaces in the B horizon. A light-colored horizon of carbonate accumulation usually occurs below the brownish B horizon.

Orthic Brown Chernozems are identified by the following properties:

1. They have a chernozemic A horizon with a color value darker than 3.5 moist and 4.5–5.5 dry.

2. They have a B horizon (Bm, Btj, or Bt) at least 5 cm thick that contains no alkaline earth carbonates.

3. They lack an Ae horizon at least 2 cm thick.

4. They lack a Bnjtj horizon or a similar horizon characteristic of intergrades to the Solonetzic order.

5. They lack evidence of gleying as indicated by faint to distinct mottling within 50 cm of the mineral surface.

Orthic Brown Chernozems and all other subgroups of Chernozemic soils may have any of the following features that are separated either at the family or series level taxonomically or as phases of any taxonomic level above the family: lithic, saline, and carbonated.

Rego Brown Chernozem

Common horizon sequence: Ah, C or Cca, or Ck

These soils have the general properties specified for the Chernozemic order and the Brown Chernozem great group. They differ from Orthic Brown Chernozems either in lacking a B horizon or in having a B horizon (Bm) less than 5 cm thick. Usually Rego Brown Chernozems have an AC profile. They may also have saline features.

Calcareous Brown Chernozem

Common horizon sequence: Ah, Bmk, Cca or Ck

These soils have the general properties of the Chernozemic order and the Brown Chernozem great group. They differ from Orthic Brown Chernozems by having a B horizon from which primary alkaline earth carbonates have not been removed

completely (Bmk). Otherwise, they have the general properties of Orthic Brown Chernozems.

Eluviated Brown Chernozem

Common horizon sequence: Ah, Ae, Btj or Bt, Cca or Ck

These soils have the general properties of the Chernozemic order and the Brown Chernozem great group. They differ from Orthic Brown Chernozems by having an eluvial horizon or horizons at least 2 cm thick (Ahe, Ae, Aej) usually underlain by a weakly to moderately developed illuvial Btj or Bt horizon. Commonly, the eluvial horizon is slightly to moderately acid. Two variations of this subgroup may be recognized.

One variation has pedons in which eluviation is indicated by light-colored, relic, macroprismatic structure below the Ah or Ap. The prismatic units break into coarse to medium platy peds that commonly have vesicular or tubular voids. This structure suggests the development of an eluvial horizon from a former prismatic B. The Bt or Btj horizon underlying the eluvial horizon usually has well-developed macroprismatic structure that breaks to blocky peds of lower color value and slightly higher chroma than the eluvial horizon.

The other variation has cumulic pedons in which the periodic deposition of transported soil materials modifies the normal horizon differentiation caused by leaching. These pedons have thick horizons of partly leached, accumulated materials overlying former A or transitional AB horizons. They are usually found on lower concave slopes where sediments are deposited from upslope. Commonly, such pedons can be recognized by their thick Ahe horizons.

Solonetzic Brown Chernozem

Common horizon sequence: Ah, Ae, Btnj or Btjnj, Csa or Ck

These soils have the general properties specified for the Chernozemic order and the Brown Chernozem great group. They differ from Orthic Brown Chernozems by having properties that indicate intergrading to the Solonetzic order. Specifically, they have a Bnj, Btjnj, or Btnj horizon and may have Ae, AB, and saline C horizons. The B horizon usually has prismatic structure and hard consistence. The prisms usually break to blocky secondary structural units having shiny, dark coatings. The eluvial horizon that is found in some Solonetzic Brown Chernozems may be similar to that described for the Eluviated Brown Chernozem subgroup. Solonetzic Brown Chernozems are usually associated with saline materials, and their B horizons usually have a higher proportion of exchangeable Na or Na and Mg than is characteristic of Orthic Brown Chernozems.

Vertic Brown Chernozem

Common horizon sequence: Ah or Ahk, Ae or Aej, Bm or Bmk, Btj or Bnjtj, Bvj, Bss or Bkss or Ckss, Ck

These soils have the general properties specified for the Chernozemic order and the Brown Chernozem great group. They may have any horizons in the upper solum common to other subgroups within the Brown Chernozem great group. However, they differ from other subgroups by having properties indicative of intergrading to the Vertisolic order. Specifically, they have a slickenside horizon (Bss, Bkss, or Ckss), the upper boundary of which occurs within 1 m of the mineral soil surface, and may have a weak vertic horizon (Bvj or BCvj).

Gleyed Brown Chernozem

Common horizon sequence: Ah, Bmgj, Ckgj

These soils have the general properties specified for the Chernozemic order and the Brown Chernozem great group. They differ from Orthic Brown Chernozems by having faint to distinct mottles indicative of gleying within 50 cm of the mineral surface.

Gleyed Rego Brown Chernozem

Common horizon sequence: Ah, Ckgj

These soils have the general properties specified for the Chernozemic order and the Brown Chernozem great group. They differ from Rego Brown Chernozems by having mottles indicative of gleying. Gleyed Rego Brown Chernozems lack a distinct B horizon at least 5 cm thick and have faint to distinct mottles within 50 cm of the mineral surface.

Gleyed Calcareous Brown Chernozem

Common horizon sequence: Ah, Bmkgj, Ckgj

These soils have the general properties specified for the Chernozemic order and the Brown Chernozem great group. They differ from Calcareous Brown Chernozems by having mottles indicative of gleying. Gleyed Calcareous Brown Chernozems have a Bmk horizon at least 5 cm thick and faint to distinct mottles within 50 cm of the mineral surface.

Gleyed Eluviated Brown Chernozem

Common horizon sequence: Ah, Aej, Btjgj or Btgj, Ckgj

These soils have the general properties specified for the Chernozemic order and the Brown Chernozem great group. They differ from Eluviated Brown Chernozems by having mottles indicative of gleying. Gleyed Eluviated Brown Chernozems have an eluvial horizon and faint to distinct mottles within 50 cm of the mineral surface.

Gleyed Solonetzic Brown Chernozem

Common horizon sequence: Ah, Ae, Bnjtjgj, Csagj

These soils have the general properties specified for the Chernozemic order and the Brown Chernozem great group. They differ from Solonetzic Brown Chernozems by having mottles indicative of gleying. Gleyed Solonetzic Brown Chernozems have a Bnj or a Btnj horizon and faint to distinct mottles within 50 cm of the mineral surface.

Gleyed Vertic Brown Chernozem

Common horizon sequence: Ah or Ahk, Ae or Aej, Bmgj or Bmkgj, Btjgj or Bnjtjgj, Bgjvj, Bgjss or Bkgjss or Ckgjss, Ckgj or Ckg

These soils have the general properties specified for the Chernozemic order and the Brown Chernozem great group. They also have properties that indicate intergrading to the Vertisolic order. Specifically, they have a slickenside horizon (Bgjss, Bkgjss, or Ckgjss), the upper boundary of which occurs within 1 m of the mineral soil surface, and may have a weak vertic horizon (Bgjvj). They differ from Vertic Brown Chernozems by having faint to distinct mottles, indicative of gleying, within 50 cm of the mineral surface.

Dark Brown Chernozem

These Chernozemic soils have A horizons somewhat darker in color than soils of the Brown Chernozem great group. Dark Brown Chernozems usually occur in association with a native vegetation of mesophytic grasses and forbs in a semiarid climate. In virgin areas the Ah horizon is usually darkest at the surface and becomes progressively lighter in color with depth.

Dark Brown Chernozems have the characteristics specified for the Chernozemic order. The A horizon has a color value darker than 3.5 moist and 3.5–4.5 dry; the chroma is usually greater than 1.5 dry. The soil climate of this great group is typically cold, rarely mild, and semiarid.

Orthic Dark Brown Chernozem

Common horizon sequence: Ah, Bm, Cca or Ck

The Orthic Dark Brown Chernozem subgroup may be thought of as the central concept of Dark Brown Chernozems. It encompasses the properties specified for the Chernozemic order and the Dark Brown Chernozem great group. Usually, Orthic Dark Brown Chernozems have brownish-colored, prismatic B horizons, and light-colored horizons of carbonate accumulation similar to those of Orthic Brown Chernozems.

Orthic Dark Brown Chernozems are identified by the following properties:

1. They have a chernozemic A horizon with a color value darker than 3.5 moist and between 3.5–4.5 dry.
2. They have a B horizon (Bm, Btj, Bt) at least 5 cm thick that does not contain alkaline earth carbonates.
3. They lack an Ae horizon at least 2 cm thick.
4. They lack a Bnjtj horizon or a similar horizon characteristic of intergrades to the Solonetzic order.
5. They lack evidence of gleying as indicated by faint to distinct mottling within 50 cm of the mineral surface.

Rego Dark Brown Chernozem

Common horizon sequence: Ah, C, Cca or Ck

These soils have the general properties specified for the Chernozemic order and the Dark Brown Chernozem great group. They differ from Orthic Dark Brown Chernozems either in lacking a B horizon or in having a B horizon (Bm) less than 5 cm thick. Usually Rego Dark Brown Chernozems have an AC profile. They may also have saline features.

Calcareous Dark Brown Chernozem

Common horizon sequence: Ah, Bmk, Cca or Ck

These soils have the general properties specified for the Chernozemic order and the Dark Brown Chernozem great group. They differ from Orthic Dark Brown Chernozems by having a B horizon from which primary alkaline earth carbonates have not been removed completely (Bmk). Otherwise, they have the general properties of Orthic Dark Brown Chernozems.

Eluviated Dark Brown Chernozem

Common horizon sequence: Ah, Ae, Btj or Bt, Cca, Ck

These soils have the general properties specified for the Chernozemic order and the Dark Brown Chernozem great group. They differ from Orthic Dark Brown Chernozems by having an eluvial horizon or horizons (Ahe, Ae, Aej) at least 2 cm thick usually underlain by a weakly to moderately developed illuvial Btj or Bt horizon. Two variations of this subgroup may be recognized. The first has an eluvial horizon developed in a former B horizon. The second has an eluvial horizon developing in an A horizon subject to periodic deposition of sediment. These variations are described in more detail for the Eluviated Brown Chernozem subgroup.

Solonetzic Dark Brown Chernozem

Common horizon sequence: Ah, Ae, Btnj or Btjnj, Csa or Ck

These soils have the general properties specified for the Chernozemic order and the Dark Brown Chernozem great group. They differ from Orthic Dark Brown Chernozems by having properties indicative of intergrading to the Solonetzic order. Specifically, they have a Bnj, Btjnj, or Btnj horizon and may have Ae, AB, and saline C horizons. The B horizon usually has prismatic structure with hard consistence. The prisms usually break to blocky secondary structural units having shiny, dark coatings. The eluvial horizon that occurs in some Solonetzic Dark Brown Chernozems may be similar to that described for the Eluviated Dark Brown Chernozem subgroup. Solonetzic Dark Brown Chernozems are usually associated with saline materials. Thus their B horizons usually have a higher proportion of exchangeable Na or Na and Mg than is characteristic of Orthic Dark Brown Chernozems.

Vertic Dark Brown Chernozem

Common horizon sequence: Ah or Ahk, Ae or Aej, Bm or Bmk, Btj or Bnjtj, Bvj, Bss or Bkss or Ckss, Ck

These soils have the general properties specified for the Chernozemic order and the Dark Brown Chernozem great group. They may have any horizons in the upper solum common to other subgroups within the Dark Brown Chernozem great group. However, they differ from other subgroups by having properties indicative of intergrading to the Vertisolic order. Specifically, they have a slickenside horizon (Bss, Bkss or Ckss), the upper boundary of which occurs within 1 m of the mineral soil surface. As well, they may have a weak vertic horizon (Bvj or BCvj).

Gleyed Dark Brown Chernozem

Common horizon sequence: Ah, Bmgj, Cgj or Ckgj

These soils have the general properties specified for the Chernozemic order and the Dark Brown Chernozem great group. They differ from Orthic Dark Brown Chernozems by having faint to distinct mottles that indicate gleying within 50 cm of the mineral surface.

Gleyed Rego Dark Brown Chernozem

Common horizon sequence: Ah, Ckgj

These soils have the general properties specified for the Chernozemic order and the Dark Brown Chernozem great group. They differ from Rego Dark Brown Chernozems by having mottles indicative of gleying. Gleyed Rego Dark Brown Chernozems lack a B horizon at least 5 cm thick and have faint to distinct mottles within 50 cm of the mineral surface.

Gleyed Calcareous Dark Brown Chernozem

Common horizon sequence: Ah, Bmkgj, Ckgj

These soils have the general properties specified for the Chernozemic order and the Dark Brown Chernozem great group. They differ from Calcareous Dark Brown Chernozems by having mottles indicative of gleying. Gleyed Calcareous Dark Brown Chernozems have a Bmk horizon at least 5 cm thick and faint to distinct mottles within 50 cm of the mineral surface.

Gleyed Eluviated Dark Brown Chernozem

Common horizon sequence: Ah, Aej, Btjgj, Cgj or Ckgj

These soils have the general properties specified for the Chernozemic order and the Dark Brown Chernozem great group. They differ from Eluviated Dark Brown Chernozems by having mottles indicative of gleying. Gleyed Eluviated Dark Brown Chernozems have an eluvial horizon and faint to distinct mottles within 50 cm of the mineral surface.

Gleyed Solonetzic Dark Brown Chernozem

Common horizon sequence: Ah, Ae, Bnjtjgj, Ckgj, Csagj

These soils have the general properties specified for the Chernozemic order and the Dark Brown Chernozem great group. They differ from Solonetzic Dark Brown Chernozems by having mottles indicative of gleying. Gleyed Solonetzic Dark Brown Chernozems have a Bnj or a Btnj horizon and faint to distinct mottles within 50 cm of the mineral surface.

Gleyed Vertic Dark Brown Chernozem

Common horizon sequence: Ah or Ahk, Ae or Aej, Bmgj or Bmkgj, Btjgj or Bnjtjgj, Bgjvj, Bgjss or Bkgjss or Ckgjss, Ckgj or Ckg

These soils have the general properties specified for the Chernozemic order and the Dark Brown Chernozem great group. In addition they have properties indicative of intergrading to the Vertisolic order. Specifically, they have a slickenside horizon (Bgjss, Bkgjss or Ckgjss), the upper boundary of which occurs within 1 m of the mineral soil surface. As well, they may have a weak vertic horizon (Bgjvj). They differ from Vertic Dark Brown Chernozems by having faint to distinct mottles, indicative of gleying, within 50 cm of the mineral surface.

Black Chernozem

These Chernozemic soils have A horizons darker in color and commonly thicker than soils of the Brown Chernozem and Dark Brown Chernozem great groups. Black Chernozems usually occur in association with a native vegetation of mesophytic grasses and forbs or with mixed grass, forb, and tree cover. Some Black Chernozems occur under alpine grass and shrub vegetation.

Black Chernozems have the characteristics specified for the order, and a chernozemic A horizon with a color value darker than 3.5 moist and dry. The chroma of the chernozemic A is usually 1.5 or less, dry. The soil climate of this great group is typically cold, rarely mild, and is subhumid.

Orthic Black Chernozem

Common horizon sequence: <u>Ah</u>, <u>Bm</u>, Cca or Ck

The Orthic Black Chernozem subgroup may be thought of as the central concept of Black Chernozems. It encompasses the properties specified for the Chernozemic order and the Black Chernozem great group. Usually Orthic Black Chernozems have brownish-colored, prismatic B horizons, and light-colored horizons of carbonate accumulation similar to those of Orthic Brown Chernozems.

Orthic Black Chernozems are identified by the following properties:

1. They have a chernozemic A horizon with a color value darker than 3.5 moist and dry.
2. They have a B horizon (Bm, Btj, Bt) at least 5 cm thick that does not contain alkaline earth carbonates.
3. They lack an eluvial horizon (Ahe, Ae, Aej) at least 2 cm thick.
4. They lack a Bnjtj or similar horizon characteristic of intergrades to the Solonetzic order.
5. They lack evidence of gleying as indicated by faint to distinct mottling within 50 cm of the mineral surface.

Rego Black Chernozem

Common horizon sequence: <u>Ah</u>, Cca or Ck

These soils have the general properties specified for the Chernozemic order and the Black Chernozem great group. They differ from Orthic Black Chernozems either in lacking a B horizon or in having a B horizon (Bm) less than 5 cm thick. Usually Rego Black Chernozems have an AC profile. They may also have saline features.

Calcareous Black Chernozem

Common horizon sequence: <u>Ah</u>, <u>Bmk</u>, Cca or Ck

These soils have the general properties of the Chernozemic order and the Black Chernozem great group. They differ from Orthic Black Chernozems by having a B horizon from which primary alkaline earth carbonates have not been removed completely (Bmk). Otherwise they have the general properties of Orthic Black Chernozems.

Eluviated Black Chernozem

Common horizon sequence: <u>Ah</u>, <u>Ae</u>, <u>Btj</u> or <u>Bt</u>, Cca or Ck

These soils have the general properties of the Chernozemic order and the Black Chernozem great group. They differ from Orthic Black Chernozems by having an eluvial horizon or horizons at least 2 cm thick (Ahe, Ae, Aej) usually underlain by a weakly to moderately developed illuvial Btj or Bt horizon.

Solonetzic Black Chernozem

Common horizon sequence: <u>Ah</u>, Ae, <u>Btnj</u> or <u>Btjnj</u>, Cs or Ck

These soils have the general properties specified for the Chernozemic order and the Black Chernozem great group. They differ from Orthic Black Chernozems by having properties indicative of intergrading to the Solonetzic order. Specifically, they have a Bnj, Btjnj, or Btnj horizon and may have Ae, AB, and saline C horizons. The B horizon usually has prismatic structure and hard consistence. The prisms usually break to blocky secondary structural units with shiny, dark coatings. The eluvial horizon that occurs in some Solonetzic Black Chernozems may be similar to that described for the Eluviated Black Chernozem subgroup. Solonetzic Black Chernozems are usually associated with saline materials. Thus, their B horizons usually have a higher proportion of exchangeable Na or Na and Mg than is characteristic of Orthic Black Chernozems.

Vertic Black Chernozem

Common horizon sequence: <u>Ah</u> or <u>Ahk</u>, Ae or Aej, Bm or Bmk, Btj or Bnjtj, Bvj, <u>Bss</u> or <u>Bkss</u> or <u>Ckss</u>, Ck

These soils have the general properties specified for the Chernozemic order and the Black Chernozem great group. They may have any horizons in the upper solum common to other subgroups within the Black Chernozem great group. However, they differ from other subgroups by having properties indicative of intergrading to the Vertisolic order. Specifically, they have a slickenside horizon (Bss, Bkss or Ckss), the upper boundary of which occurs within 1 m of the mineral soil surface. As well, they may have a weak vertic horizon (Bvj or BCvj).

Gleyed Black Chernozem

Common horizon sequence: <u>Ah</u>, <u>Bmgj</u>, Ckgj

These soils have the general properties specified for the Chernozemic order and the Black Chernozem great group. They differ from Orthic Black Chernozems by having faint to distinct mottles indicative of gleying within 50 cm of the mineral surface.

Gleyed Rego Black Chernozem

Common horizon sequence: <u>Ah</u>, <u>Ckgj</u>

These soils have the general properties specified for the Chernozemic order and the Black Chernozem great group. They differ from Rego Black Chernozems by having mottles indicative of gleying. Gleyed Rego Black Chernozems lack a B horizon at least 5 cm thick and have faint to distinct mottles within 50 cm of the mineral surface.

Gleyed Calcareous Black Chernozem

Common horizon sequence: <u>Ah</u>, <u>Bmkgj</u>, Ckgj

These soils have the general properties specified for the Chernozemic order and the Black Chernozem great group. They differ from Calcareous Black Chernozems by having mottles indicative of gleying. Gleyed Calcareous Black Chernozems have a Bmk horizon at least 5 cm thick and faint to distinct mottles within 50 cm of the mineral surface.

Gleyed Eluviated Black Chernozem

Common horizon sequence: <u>Ah</u>, <u>Aej</u>, <u>Btjgj</u>, Ckgj

These soils have the general properties specified for the Chernozemic order and the Black Chernozem great group. They differ from Eluviated Black Chernozems by having mottles indicative of gleying. Gleyed Eluviated Black Chernozems have an eluvial horizon and faint to distinct mottles within 50 cm of the mineral surface.

Gleyed Solonetzic Black Chernozem

Common horizon sequence: <u>Ah</u>, Ae, <u>Bnjtjgj</u>, Ckgj, Csgj

These soils have the general properties specified for the Chernozemic order and the Black Chernozem great group. They differ from Solonetzic Black Chernozems by having mottles indicative of gleying. Gleyed Solonetzic Black Chernozems have a Bnj or a Btnj horizon and faint to distinct mottling within 50 cm of the mineral surface.

Gleyed Vertic Black Chernozem

Common horizon sequence: <u>Ah</u> or <u>Ahk</u>, Ae or Aej, Bmgj or Bmkgj, Btjgj or Bnjtjgj, Bgjvj, <u>Bgjss</u> or <u>Bkgjss</u> or <u>Ckgjss</u>, Ckgj or Ckg

These soils have the general properties specified for the Chernozemic order and the Black Chernozem great group. They also have properties indicative of intergrading to the Vertisolic order. Specifically, they have a slickenside horizon (Bgjss, Bkgjss or Ckgjss), the upper boundary of which occurs within 1 m of the mineral soil surface. They may have a weak vertic horizon (Bgjvj). As well, they differ from Vertic Black Chernozems by having faint to distinct mottles, indicative of gleying, within 50 cm of the mineral surface.

Dark Gray Chernozem

These soils have chernozemic A horizons with characteristics indicative of eluviation associated with soils developed under forest vegetation. Dark Gray Chernozems usually occur under mixed native vegetation of trees, shrubs, forbs, and grasses in forest–grassland transition zones in areas of cold, subhumid soil climate. Virgin Dark Gray Chernozems usually have leaf mats (L-H horizons) overlying Ah or Ahe horizons. The peds of the A horizons may have dark-colored surfaces, but the crushed peds usually have gray or brownish colors of higher value or chroma. A "salt and pepper" effect, light gray spots, or bands on a darker matrix may be observable in the A horizons, which tend to have platy structure.

Dark Gray Chernozems have the characteristics specified for the order. Virgin soils have a chernozemic A horizon with a color value darker than 3.5 moist and 4.5 dry. An Ap horizon 10 cm thick must have a color value darker than 3.5 moist and 5 dry. The chroma of the A horizon is usually 1.5 or less, dry. Dark Gray Chernozems may have a light-colored Ae horizon 5 cm or less thick provided that the mixed surface horizon (Ap) meets the requirements specified for the great group. Similar soils having a distinct Ae horizon thicker than 5 cm below the chernozemic A horizon and a Bt horizon are classified as Dark Gray Luvisols.

Orthic Dark Gray Chernozem

Common horizon sequence: <u>Ahe</u>, Ae, <u>Bm</u> or <u>Btj</u> or <u>Bt</u>, Cca or Ck

The Orthic Dark Gray Chernozem subgroup may be thought of as the central concept of Dark Gray Chernozems. It encompasses the properties specified for the Chernozemic order and the Dark Gray Chernozem great group. Usually Orthic Dark Gray Chernozems have brownish-colored, prismatic B horizons that commonly meet the requirements of a Bt horizon.

Orthic Dark Gray Chernozems are identified by the following properties:

1. They have a chernozemic A horizon with a color value darker than 3.5 moist and 3.5–4.5 dry for virgin soils (3.5–5 dry for Ap).

2. They have a B horizon (Bm, Btj, Bt) at least 5 cm thick that does not contain alkaline earth carbonates.

3. They lack an Ae horizon more than 5 cm thick.

4. They lack a Bnjtj horizon or a similar horizon characteristic of intergrades to the Solonetzic order.

5. They lack evidence of gleying as indicated by faint to distinct mottling within 50 cm of the mineral surface.

Rego Dark Gray Chernozem

Common horizon sequence: <u>Ah</u> or <u>Ahe</u>, Cca or Ck

These soils have the general properties specified for the Chernozemic order and the Dark Gray Chernozem great group. They differ from Orthic Dark Gray Chernozems either in lacking a B horizon or in having a B horizon (Bm) less than 5 cm thick. Usually Rego Dark Gray Chernozems have an AC profile. They may also have saline features.

Calcareous Dark Gray Chernozem

Common horizon sequence: <u>Ah</u> or <u>Ahe</u>, <u>Bmk</u>, Cca or Ck

These soils have the general properties of the Chernozemic order and the Dark Gray Chernozem great group. They differ from Orthic Dark Gray Chernozems by having a B horizon from which primary alkaline earth carbonates have not been removed completely (Bmk). Otherwise they have the general properties of Orthic Dark Gray Chernozems.

Solonetzic Dark Gray Chernozem

Common horizon sequence: <u>Ahe</u>, Ae, <u>Bnjtj</u>, Cs or Ck

These soils have the general properties specified for the Chernozemic order and the Dark Gray Chernozem great group. They differ from Orthic Dark Gray Chernozems by having properties indicative of intergrading to the Solonetzic order. Specifically, they have either a Bnj, Bnjtj, or Btnj horizon, and they may have Ae, AB, and saline C horizons. The Ae horizon, if present, is not more than 5 cm thick. The B horizon usually has prismatic structure and hard consistence, and the prisms usually break to blocky secondary structural units with shiny, dark coatings. Solonetzic Dark Gray Chernozems are usually associated with saline materials. Thus, their B horizons usually have a higher proportion of exchangeable Na or Na and Mg than is characteristic of Orthic Dark Gray Chernozems.

Vertic Dark Gray Chernozem

Common horizon sequence: <u>Ah</u> or <u>Ahk</u>, Ae or Aej, Bm or Bmk, Btj or Bnjtj, Bvj, <u>Bss</u> or <u>Bkss</u> or <u>Ckss</u>, Ck

These soils have the general properties specified for the Chernozemic order and the Dark Gray Chernozem great group. They may have any horizons in the upper solum common to other subgroups within the Dark Gray Chernozem great group. However, they differ from other subgroups by having properties indicative of intergrading to the Vertisolic order. Specifically, they have a slickenside horizon (Bss, Bkss or Ckss), the upper boundary of which occurs within 1 m of the mineral soil surface. As well, they may have a weak vertic horizon (Bvj or BCvj).

Gleyed Dark Gray Chernozem

Common horizon sequence: <u>Ahe</u>, <u>Bmgj</u>, Ckgj

These soils have the general properties specified for the Chernozemic order and the Dark Gray Chernozem great group. They differ from Orthic Dark Gray Chernozems by having faint to distinct mottles indicative of gleying within 50 cm of the mineral surface.

Gleyed Rego Dark Gray Chernozem

Common horizon sequence: <u>Ahe</u>, <u>Ckgj</u>

These soils have the general properties specified for the Chernozemic order and the Dark Gray Chernozem great group. They differ from Rego Dark Gray Chernozems by having mottles indicative of gleying. Gleyed Rego Dark Gray Chernozems lack a B horizon at least 5 cm thick and have faint to distinct mottles within 50 cm of the mineral surface.

Gleyed Calcareous Dark Gray Chernozem

Common horizon sequence: <u>Ahe</u>, <u>Bmkgj</u>, Ckgj

These soils have the general properties specified for the Chernozemic order and the Dark Gray Chernozem great group. They differ from Calcareous Dark Gray Chernozems by having mottles indicative of gleying. Gleyed Calcareous Dark Gray Chernozems have a Bmk horizon at least 5 cm thick and faint to distinct mottles within 50 cm of the mineral surface.

Gleyed Solonetzic Dark Gray Chernozem

Common horizon sequence: <u>Ahe</u>, Ae, <u>Bnjtjgj</u>, Ckgj, Csgj

These soils have the general properties specified for the Chernozemic order and the Dark Gray Chernozem great group. They differ from Solonetzic Dark Gray Chernozems by having mottles indicative of gleying. Gleyed Solonetzic Dark Gray Chernozems have a Bnj or a Btnj horizon and faint to distinct mottles within 50 cm of the mineral surface.

Gleyed Vertic Dark Gray Chernozem

Common horizon sequence: <u>Ah</u> or <u>Ahk</u>, Ae or Aej, Bmgj or Bmkgj, Btjgj or Bnjtjgj, Bgjvj, <u>Bgjss</u> or <u>Bkgjss</u> or <u>Ckgjss</u>, Ckgj or Ckg

These soils have the general properties specified for the Chernozemic order and the Dark Gray Chernozem great group. They also have properties indicative of intergrading to the Vertisolic order. Specifically, they have a slickenside horizon (Bgjss, Bkgjss or Ckgjss), the upper boundary of which occurs within 1 m of the mineral soil surface. As well, they may have a weak vertic horizon (Bgjvj). They differ from Vertic Dark Gray Chernozems by having faint to distinct mottles, indicative of gleying, within 50 cm of the mineral surface.

Chapter 6

Cryosolic Order

Great Group	*Subgroup*
Turbic Cryosol	Orthic Eutric Turbic Cryosol OE.TC
	Orthic Dystric Turbic Cryosol OD.TC
	Brunisolic Eutric Turbic Cryosol BRE.TC
	Brunisolic Dystric Turbic Cryosol BRD.TC
	Gleysolic Turbic Cryosol GL.TC
	Regosolic Turbic Cryosol R.TC
	Histic Eutric Turbic Cryosol HE.TC
	Histic Dystric Turbic Cryosol HD.TC
	Histic Regosolic Turbic Cryosol HR.TC
Static Cryosol	Orthic Eutric Static Cryosol OE.SC
	Orthic Dystric Static Cryosol OD.SC
	Brunisolic Eutric Static Cryosol BRE.SC
	Brunisolic Dystric Static Cryosol BRD.SC
	Luvisolic Static Cryosol L.SC
	Gleysolic Static Cryosol GL.SC
	Regosolic Static Cryosol R.SC
	Histic Eutric Static Cryosol HE.SC
	Histic Dystric Static Cryosol HD.SC
	Histic Regosolic Static Cryosol HR.SC
Organic Cryosol	Fibric Organic Cryosol FI.OC
	Mesic Organic Cryosol ME.OC
	Humic Organic Cryosol HU.OC
	Terric Fibric Organic Cryosol TFI.OC
	Terric Mesic Organic Cryosol TME.OC
	Terric Humic Organic Cryosol THU.OC
	Glacic Organic Cryosol GC.OC

A diagrammatic representation of profiles of some subgroups of the Cryosolic order is shown in Figures 30 and 31. Individual subgroups may include soils that have horizon sequences different from those shown. In the description of each subgroup, presented later in this chapter, a common horizon sequence is given; diagnostic horizons are underlined and some other commonly occurring horizons are listed.

Soils of the Cryosolic order occupy much of the northern third of Canada where permafrost exists close to the surface of both mineral and organic deposits. Cryosolic soils predominate north of the tree line, are common in the subarctic forest area in fine-textured soils, and extend into the boreal forest in some organic materials and into some alpine areas of mountainous regions. Cryoturbation of these soils is common and may be indicated by patterned ground features such as sorted and nonsorted nets, circles, polygons, stripes, and earth hummocks.

Cryosolic soils are formed in either mineral or organic materials that have permafrost either within 1 m of the surface or within 2 m if the pedon has been strongly cryoturbated laterally within the active layer, as indicated by disrupted, mixed, or broken horizons. They have a mean annual temperature ≤0°C. Differentiation of Cryosolic soils from soils of other orders involves either determining or estimating the depth to permafrost.

The Cryosolic order is divided into three great groups: Turbic Cryosol, Static Cryosol, and Organic Cryosol based on the degree of cryoturbation and the nature of soil material, mineral or organic, as indicated in the Cryosolic order chart.

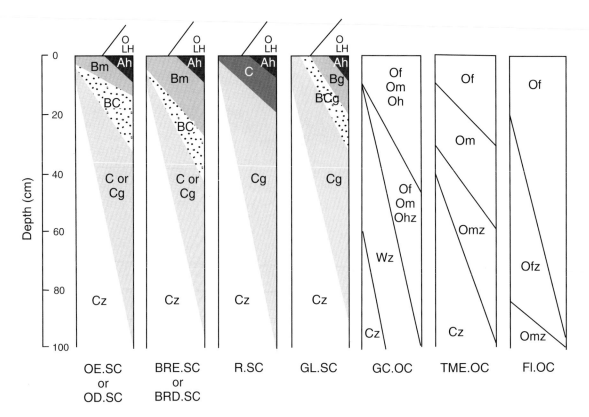

Figure 30 Diagrammatic horizon pattern of some subgroups of the Cryosolic order.

Cryosolic Order

	Turbic Cryosol	Static Cryosol	Organic Cryosol
Soil	mineral	mineral	organic
Cryoturbation	marked, usually patterned ground	none	none
Permafrost	within 2 m of surface	within 1 m of surface	within 1 m of surface

Turbic Cryosol

These are Cryosolic soils that have developed primarily in mineral material and have marked evidence of cryoturbation (Harris et al. 1988). They generally occur on patterned ground, which includes such cryogenic forms as sorted and nonsorted circles, nets, polygons, stripes, and steps in stony or coarse-textured material and nonsorted units such as earth hummocks in medium- and fine-textured materials. The pedon includes all elements of the microtopography in cycles less than 7 m in width. Processes include sorting of different-sized particles and mixing of both mineral and organic material from different horizons. Organic (Oy) or organic-rich, mineral horizons (Ahy) are characteristically present in the region of the permafrost table (upper surface of permafrost

layer), and there is generally a buildup of ice in the upper part of the permafrost layer.

Turbic Cryosols are mineral soils that have permafrost within 2 m of the surface and show marked evidence of cryoturbation laterally within the active layer, as indicated by disrupted or mixed or broken horizons, or displaced material or a combination of both.

Orthic Eutric Turbic Cryosol

Common horizon sequence: Om, <u>Bmy</u> , BCy, Cgy, Omy, <u>Cz</u>

These Turbic Cryosols have a relatively high degree of base saturation, as indicated by their pH. They are identified by the following properties:

1. These soils have a Bmy horizon and may have a Bm horizon less than 10 cm thick.

2. The horizons are strongly disrupted by cryoturbation. Tongues of mineral and organic horizons, organic and mineral intrusions, and oriented stones commonly occur.

3. These soils have a pH (0.01 M CaCl₂) of 5.5 or greater in some or all of the B horizons.

4. The surface horizons are not strongly gleyed, but there is usually a gleyed horizon immediately above the permafrost table.

Orthic Eutric Turbic Cryosols have a Bmy horizon and subsurface organic (Omy, Ohy) or organic-rich, mineral horizons (Ahy). Surface organic horizons up to 15 cm thick, moder Ah horizons, and Bm horizons less than 10 cm thick may occur.

Orthic Dystric Turbic Cryosol

Common horizon sequence: Om, <u>Bmy</u>, BCy, Cgy, Omy, <u>Cz</u>

These acidic Turbic Cryosols have a low degree of base saturation, as indicated by their pH. They are identified by the following properties:

1. These soils have a Bmy horizon and may have a Bm horizon less than 10 cm thick.

2. The horizons are strongly disrupted by cryoturbation. Tongues of mineral and organic horizons, organic and mineral intrusions, and oriented stones commonly occur.

3. These soils have a pH (0.01 M CaCl₂) of less than 5.5 throughout the B horizons.

4. The surface horizons are not strongly gleyed, but there is usually a gleyed horizon immediately above the permafrost table.

Orthic Dystric Turbic Cryosols have a Bmy horizon and subsurface organic (Omy, Ohy) or organic-rich, mineral horizons (Ahy). Surface organic horizons up to 15 cm thick, moder Ah horizons, and Bm horizons less than 10 cm thick may occur.

Brunisolic Eutric Turbic Cryosol

Common horizon sequence: Om, <u>Bm</u>, Bmy or BCy, Cgy, Omy, <u>Cz</u>

These Turbic Cryosols have a relatively high degree of base saturation, as indicated by their pH. They are identified by the following properties:

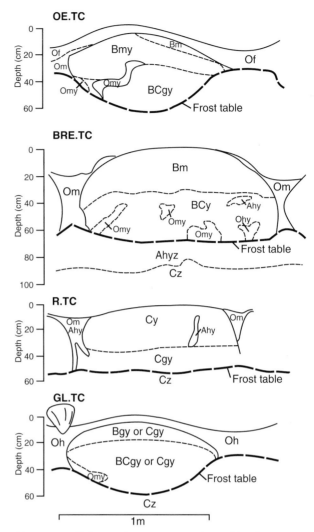

Figure 31 Schematic representation of horizon pattern in some Turbic subgroups of the Cryosolic order.

1. These soils have a Bm horizon, at least 10 cm thick, which is continuous over the imperfectly to well-drained part of the pedon that is relatively unaffected by cryoturbation.

2. The horizons, other than the Bm, are strongly disrupted by cryoturbation. Tongues of mineral and organic horizons, organic and mineral intrusions, and oriented stones commonly occur.

3. These soils have a pH (0.01 M CaCl₂) of 5.5 or greater in some or all of the B horizons.

4. The surface horizons are not strongly gleyed, but there is usually a gleyed horizon immediately above the permafrost table.

Brunisolic Dystric Turbic Cryosol

Common horizon sequence: Om, <u>Bm</u>, Bmy or BCy, Cgy, Omy, <u>Cz</u>

75

These acidic Turbic Cryosols have a low degree of base saturation, as indicated by their pH. They are identified by the following properties:

1. These soils have a Bm horizon, at least 10 cm thick, which is continuous over the imperfectly to well-drained part of the pedon that is relatively unaffected by cryoturbation.

2. The horizons, other than the Bm, are strongly disrupted by cryoturbation. Tongues of mineral and organic horizons, organic and mineral intrusions, and oriented stones commonly occur.

3. These soils have a pH (0.01 M $CaCl_2$) of less than 5.5 throughout the B horizons.

4. The surface horizons are not strongly gleyed, but there is usually a gleyed horizon immediately above the permafrost table.

Gleysolic Turbic Cryosol

Common horizon sequence: Om, Bgy or Cgy (or both), Cz

These Turbic Cryosols have developed in poorly drained areas under reducing conditions. The name Gleysolic, rather than Gleyed, is used because these soils are not equivalent in degree of gleying to Gleyed subgroups of other orders. Gleysolic subgroups of Cryosolic soils have evidence of gleying similar in degree to that of soils of the Gleysolic order. They are identified by the following properties:

1. They have evidence of gleying in the form of low chromas or mottling to the mineral surface.

2. Their uppermost mineral horizon, Bgy or Cgy, may be overlain by organic layers less than 40 cm thick, or a combination of surface and subsurface horizons >15 cm thick.

Regosolic Turbic Cryosol

Common horizon sequence: Om, Cy, Cgy, Cz

These Turbic Cryosols have developed on recently deposited or strongly cryoturbated soil materials. They are identified by the following properties:

1. These soils lack B horizons.

2. These soils usually have little incorporated organic matter.

3. Cryoturbation is manifested by oriented stones, displacement of materials, and sorting.

Histic Eutric Turbic Cryosol

Common horizon sequence: Om, Ah, Bmy or Bm (or both) or Cgy, Cz

These Turbic Cryosols have a relatively high degree of base saturation, as indicated by their pH, and thick (>15 cm) organic (peaty) horizons in the upper 1 m of the solum. They are identified by the following properties:

1. These soils have either a continuous surface organic horizon (Ohy, Hy) ranging in thickness from >15 to 40 cm, or a combination of surface and subsurface organic horizons >15 cm thick.

2. They have a Bm or Bmy horizon, or both, which is continuous over the imperfectly to well-drained part of the pedon.

3. The horizons, other than the Bm, are strongly disrupted by cryoturbation. Tongues of mineral and organic horizons, organic and mineral intrusions, and oriented stones commonly occur.

4. These soils have a pH (0.01 M $CaCl_2$) of 5.5 or greater in some or all of the B horizons.

5. The surface horizons are not strongly gleyed, but there is usually a gleyed horizon immediately above the permafrost table.

Histic Dystric Turbic Cryosol

Common horizon sequence: Om, Ah, Bmy or Bm (or both) or Cgy, Cz

These acidic Turbic Cryosols have a low degree of base saturation, as indicated by their pH, and thick (>15 cm) organic (peaty) horizons in the upper 1 m of the solum. They are identified by the following properties:

1. These soils have either a continuous surface organic horizon (Ohy, Hy) ranging in thickness from >15 to 40 cm, or a combination of surface and subsurface organic horizons >15 cm thick.

2. They have a Bm or Bmy horizon (or both), which is continuous over the imperfectly to well-drained part of the pedon that is relatively unaffected by cryoturbation.

3. The horizons, other than the Bm, are strongly disrupted by cryoturbation. Tongues of mineral and organic horizons, organic and mineral intrusions, and oriented stones commonly occur.

4. These soils have a pH (0.01 M CaCl$_2$) of less than 5.5 throughout the B horizons.

5. The surface horizons are not strongly gleyed, but there is usually a gleyed horizon immediately above the permafrost table.

Histic Regosolic Turbic Cryosol

Common horizon sequence: Om, Cy, Cgy, Cz

These Turbic Cryosols have developed on recently deposited or strongly cryoturbated soil materials. They have thick (>15 cm) organic (peaty) horizons in the upper 1 m of the solum. They are identified by the following properties:

1. These soils have either a continuous surface organic horizon (Ohy, Hy) ranging in thickness from >15 to 40 cm, or a combination of surface and subsurface organic horizons >15 cm thick.

2. They lack B horizons.

3. Cryoturbation is manifested by oriented stones, displacement of materials, and sorting.

4. The surface horizons are not strongly gleyed, but there is usually a gleyed horizon immediately above the permafrost table.

Static Cryosol

These Cryosolic soils have developed primarily in coarse-textured mineral parent materials, or in a wide textural range of recently deposited or disturbed sediments where evidence of cryoturbation is still largely absent, or in both. They may have organic surface horizons less than 40 cm thick.

Static Cryosols have permafrost within 1 m of the surface but show little or no evidence of cryoturbation or features that indicate cryoturbation. They may be associated with landforms exhibiting patterned ground features such as polygons.

Orthic Eutric Static Cryosol

Common horizon sequence: Om, LFH, Bm, BCgj, Cz

These Static Cryosols have a relatively high degree of base saturation, as indicated by their pH. They are identified by the following properties:

1. These soils have a Bm horizon less than 10 cm thick.

2. They have a pH (0.01 M CaCl$_2$) of 5.5 or greater in some or all of the B horizons.

3. The surface horizons are not strongly gleyed, but gley features commonly occur immediately above the permafrost table.

Orthic Dystric Static Cryosol

Common horizon sequence: Om, LFH, Bm, BCgj, Cz

These acidic Static Cryosols have a low degree of base saturation, as indicated by their pH. They are identified by the following properties:

1. These soils have a Bm horizon less than 10 cm thick.

2. They have a pH (0.01 M CaCl$_2$) of less than 5.5 throughout the B horizons.

3. The surface horizons are not strongly gleyed, but gley features commonly occur immediately above the permafrost table.

Brunisolic Eutric Static Cryosol

Common horizon sequence: LFH, Bm, BCgj, Cz

These Static Cryosols have a relatively high degree of base saturation, as indicated by their pH. They are identified by the following properties:

1. These soils have a Bm horizon at least 10 cm thick.

2. They have a pH (0.01 M CaCl$_2$) of 5.5 or greater in some or all of the B horizons.

Brunisolic Dystric Static Cryosol

Common horizon sequence: LFH, Bm, BCgj, Cz

These acidic Static Cryosols have a low degree of base saturation, as indicated by their pH. They are identified by the following properties:

1. These soils have a Bm horizon at least 10 cm thick.

2. They have a pH (0.01 M CaCl$_2$) of less than 5.5 throughout the B horizons.

Luvisolic Static Cryosol

Common horizon sequence: LFH, Om, Ah, or Ae, Bt, Cg, Cz

These Static Cryosols have a clay accumulation in the B horizon (Bt). They occur mainly on fine-textured parent materials under forest vegetation. They are identified by the following properties:

1. These soils have an eluvial horizon (Ahe, Ae, Aeg).

2. They have a Bt horizon that is greater than 10 cm thick.

3. Some mottling may occur in the lower part of the solum.

Gleysolic Static Cryosol

Common horizon sequence: Om, Bg or Cg (or both), Cz

These Static Cryosols have developed in poorly drained areas under reducing conditions. The name Gleysolic, rather than Gleyed, is used because these soils are not equivalent in degree of gleying to Gleyed subgroups of other orders. Gleysolic subgroups of Cryosolic soils have evidence of gleying similar in degree to that of soils of the Gleysolic order. They are identified by the following properties:

1. These soils have evidence of gleying in the form of low chromas or mottling to the mineral surface.

2. Their surface mineral horizon, Bg or Cg, may be overlain by organic layers less than 40 cm thick.

Regosolic Static Cryosol

Common horizon sequence: C, Cg, Cz

These Static Cryosols have developed on recently deposited or strongly cryoturbated soil materials. They are identified by the following properties:

1. These soils lack B horizons.

2. These soils may have thin (≤15 cm thick) peaty organic layers and a thin Ah horizon.

Histic Eutric Static Cryosol

Common horizon sequence: Om, Ah, Bm, Cg, Cz

These Static Cryosols have a relatively high degree of base saturation, as indicated by their pH, and thick (>15 cm) organic (peaty) horizons in the upper 1 m of the solum. They are identified by the following properties:

1. These soils have a continuous surface organic horizon (O, H) ranging in thickness from >15 to 40 cm.

2. These soils have Bm horizons that are continuous over the imperfectly to well-drained part of the pedon.

3. These soils have a pH (0.01 M $CaCl_2$) of 5.5 or greater in some or all of the B horizons.

4. The surface horizons are not strongly gleyed, but there is usually a gleyed horizon immediately above the permafrost table.

Histic Dystric Static Cryosol

Common horizon sequence: Om, Ah, Bm, Cg, Cz

These acidic Static Cryosols have a low degree of base saturation, as indicated by their pH, and thick (>15 cm) organic (peaty) horizons in the upper 1 m of the solum. They are identified by the following properties:

1. These soils have a surface organic horizon (O, H) ranging in thickness from >15 to 40 cm.

2. These soils have Bm horizons that are continuous over the imperfectly to well-drained part of the pedon that is relatively unaffected by cryoturbation.

3. These soils have a pH (0.01 M $CaCl_2$) of less than 5.5 throughout the B horizons.

4. The surface horizons are not strongly gleyed, but there is usually a gleyed horizon immediately above the permafrost table.

Histic Regosolic Static Cryosol

Common horizon sequence: Om, C, Cg, Cz

These Static Cryosols have relatively thick (>15 cm) organic (peaty) horizons in the upper 1 m of the solum. They are identified by the following properties:

1. These soils have a surface organic horizon (O, H) ranging in thickness from >15 to 40 cm.

2. They lack B horizons.

3. They may have thin peaty organic layers and a thin Ah horizon.

4. The surface horizons are not strongly gleyed, but there is usually a gleyed horizon immediately above the permafrost table.

Organic Cryosol

These Cryosolic soils have developed primarily from organic material containing more than 17% organic carbon by weight and are underlain by permafrost within 1 m of the surface. They are greater than 40 cm thick, or greater than 10 cm thick over either a lithic contact or an ice layer that is at least 30 cm thick. Landforms associated with Organic Cryosols include palsas, peat polygons, and high-centre polygons.

Fibric Organic Cryosol

Common horizon sequence: Of or Om, Of, Ofz

These Organic Cryosols are identified by the following properties:

1. They have organic layers thicker than 1 m.

2. They are composed dominantly of fibric material in the control section below a depth of 40 cm.

Mesic Organic Cryosol

Common horizon sequence: Of or Om, Om, Omz

These Organic Cryosols are identified by the following properties:

1. They have organic layers thicker than 1 m.

2. They are composed dominantly of mesic material in the control section below a depth of 40 cm.

Humic Organic Cryosol

Common horizon sequence: Om or Oh, Ohz

These Organic Cryosols are identified by the following properties:

1. They have organic layers thicker than 1 m.

2. They are composed dominantly of humic material in the control section below a depth of 40 cm.

Terric Fibric Organic Cryosol

Common horizon sequence: Of, Ofz, Cz

These Organic Cryosols are identified by the following properties:

1. They have a mineral contact within 1 m of the surface, or the mineral layer is greater than 30 cm thick with an upper boundary within 1 m of the surface.

2. They are composed dominantly of fibric material above the mineral contact.

Terric Mesic Organic Cryosol

Common horizon sequence: Om, Omz, Cz

These Organic Cryosols are identified by the following properties:

1. They have a mineral contact within 1 m of the surface, or the mineral layer is greater than 30 cm thick with an upper boundary within 1 m of the surface.

2. They are composed dominantly of mesic material above the mineral contact.

Terric Humic Organic Cryosol

Common horizon sequence: Oh, Ohz, Cz

These Organic Cryosols are identified by the following properties:

1. They have a mineral contact within 1 m of the surface, or the mineral layer is greater than 30 cm thick with an upper boundary within 1 m of the surface.

2. They are composed dominantly of humic material above the mineral contact.

Glacic Organic Cryosol

Common horizon sequence: Of, Om or Oh, Wz

These Organic Cryosols are identified by the following properties:

1. They have a layer of ground ice greater than 30 cm thick with an upper boundary within 1 m of the surface.

2. The ice layer contains more than 95% ice by volume.

Chapter 7

Gleysolic Order

Great Group	Subgroup
Luvic Gleysol	Vertic Luvic Gleysol V.LG
	Solonetzic Luvic Gleysol SZ.LG
	Fragic Luvic Gleysol FR.LG
	Humic Luvic Gleysol HU.LG
	Fera Luvic Gleysol FE.LG
	Orthic Luvic Gleysol O.LG
Humic Gleysol	Vertic Humic Gleysol V.HG
	Solonetzic Humic Gleysol SZ.HG
	Fera Humic Gleysol FE.HG
	Orthic Humic Gleysol O.HG
	Rego Humic Gleysol R.HG
Gleysol	Vertic Gleysol V.G
	Solonetzic Gleysol SZ.G
	Fera Gleysol FE.G
	Orthic Gleysol O.G
	Rego Gleysol R.G

Note: The great groups and subgroups are arranged in the order in which they are keyed out. For example, if a Gleysolic soil has a Btg horizon it is classified as a Luvic Gleysol, regardless of whether or not it has any of the following: Ah, Bn, Bgf, or fragipan. The Luvic Gleysol is the first great group keyed out. Similarly, at the subgroup level, if a Luvic Gleysol has a solonetzic B horizon it is classified as a Solonetzic Luvic Gleysol, regardless of whether or not it has any of the following: fragipan, Ah, or Bgf. In essence, any class at the great group or subgroup level as listed does not have the diagnostic properties of classes listed above it. For example, a Rego Gleysol does not have any of the following: a B horizon as defined for Orthic Gleysol, a Bgf horizon, or a solonetzic B horizon.

A diagrammatic representation of profiles of some subgroups of the Gleysolic order is shown in Figure 32. Individual subgroups may include soils that have horizon sequences different from those shown. In the description of each subgroup, presented later in this chapter, a common horizon sequence is given; diagnostic horizons are underlined and some other commonly occurring horizons are listed.

Gleysolic soils are defined on the basis of color and mottling, which are considered to indicate the influence of periodic or sustained reducing conditions during their genesis. The criteria that follow apply to all horizons except Ah, Ap, and Ae. However, if the Ae horizon is thicker than 20 cm and its lower boundary is more than 60 cm below the mineral soil surface, the criteria do apply to the Ae. Also, if the Ah or Ap horizon is thicker than 50 cm, the color criteria apply to the mineral horizon immediately below. Apart from these exceptions the criteria are as follows: Gleysolic soils have a horizon or subhorizon, at least 10 cm thick (the upper boundary of which occurs within 50 cm of the mineral surface), with moist colors, as follows:

1. For all but red soil materials (hues of 5YR or redder where the soil color fades slowly upon treatment of the soil with dithionite).

   a. Dominant chromas of 1 or less or hues bluer than 10Y with or without mottles; or

   b. Dominant chromas of 2 or less in hues of 10YR and 7.5YR accompanied by prominent mottles 1 mm or larger in cross section and occupying at least 2% of the exposed, unsmeared 10 cm layer; or

   c. Dominant chromas of 3 or less in hues yellower than 10YR accompanied by prominent mottles 1 mm or larger in cross section and occupying at least 2% of the exposed, unsmeared 10 cm layer.

2. For red soil materials (hues of 5YR or redder where the soil color fades slowly upon treatment of the soil with dithionite).

   a. Distinct or prominent mottles at least 1 mm in diameter occupy at least 2% of the exposed, unsmeared 10 cm layer.

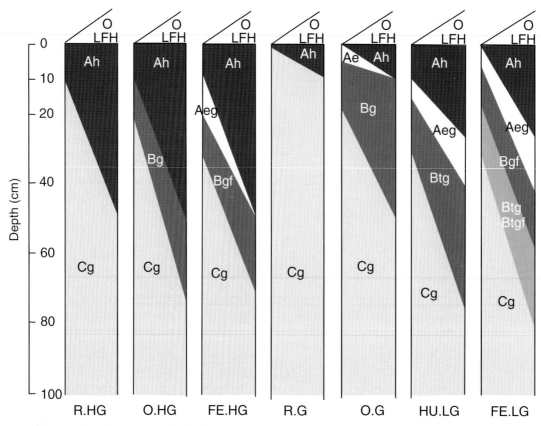

Figure 32 Diagrammatic horizon pattern of some subgroups of the Gleysolic order.

Soils of the Gleysolic order have properties that indicate prolonged periods of intermittent or continuous saturation with water and reducing conditions during their genesis. Saturation with water may result from either a high groundwater table or temporary accumulation of water above a relatively impermeable layer, or both. In contrast, soils saturated periodically with aerated water or saturated for prolonged cold periods, which restricts biological activity without developing properties associated with reducing conditions, are not classified as Gleysols.

Gleysolic soils are associated with a number of different moisture regimes that may change during the genesis of the soil. They commonly have peraquic or aquic regimes, but some have aqueous regimes and others are now rarely, if ever, saturated with water. Those that are rarely saturated now presumably had aquic moisture regimes in the past and were once under reducing conditions. Drainage, isostatic uplift, or other factors have resulted in a changed moisture regime in these soils.

Gleysolic soils occur in association with other soils in the landscape, in some cases as the dominant soils, in others as a minor component. In areas of subhumid climate, Gleysolic soils occur commonly in shallow depressions and on level lowlands that are saturated with water every spring. In more humid areas, they may also occur on slopes and on undulating terrain. The native vegetation associated with Gleysolic soils commonly differs from that of associated soils of other orders.

Some notes on the basis of the color criteria follow. The criteria are based on color because it is the most easily observable and most useful indicator of the oxidation–reduction status prevailing during soil genesis. Redox potential measured at several depths within pedons throughout the period when the soil is not frozen provides useful information on current redox conditions. Values of E **pt** of 100 mV or less are associated with reduced forms of Mn and Fe. Such values, however, indicate only present redox conditions, not those that existed over long periods during which the soil developed. Similarly, monitoring of moisture regime properties, such as depth to water table, provides valuable information on the present state of the soil, but it does not necessarily indicate the prevailing moisture regime during soil genesis.

Color criteria used for red soil materials are different from those used for material of other colors because even prolonged saturation and, presumably, reducing conditions have not been found to result in the development of drab gray colors in such materials. Usually, however, such soils are mottled in horizons near the surface. In some cases gray mottles occur in a reddish matrix, in others strong brown or yellowish-red mottles occur in a matrix of lower chroma. The dominant color is considered to be the matrix color.

Exceptions had to be made in applying the criteria to soils with Ah, Ap, or Ae horizons because chromas of 1 occur in some horizons of oxidized soils. Furthermore, prominent mottling may occur in Ae horizons overlying relatively impermeable horizons of generally oxidized soils. In the case of thick Ae horizons, however, prominent mottling in the upper part of the horizon is thought to indicate periodic reducing conditions near the surface. These exceptions have not been tested and will probably require adjustment.

The color criteria specify a minimum size and abundance of mottles in a subhorizon 10 cm thick or thicker, because it seems unreasonable to base classification at the order level on the occurrence of few or fine mottles in a thin layer. Care is required in estimating the abundance of mottles; smearing of ocherous material on the profile can result in overestimates, and failure to look for both inped and exped mottles can result in underestimates. The use of mottle charts facilitates estimates of abundance.

Distinguishing Gleysolic Soils from Soils of Other Orders

Listed below are guidelines for distinguishing Gleysolic soils from soils of other orders with which they might be confused.

Brunisolic Gleyed subgroups of Brunisolic soils are differentiated from Gleysolic soils on the evidence that gleying is too weakly expressed to meet the specifications of Gleysolic soils.

Chernozemic Some soils have a chernozemic A horizon and dull colors or mottling, indicating gleying within the control section. Those meeting the requirements specified for Gleysolic soils are classified in the Gleysolic order. Gleyed subgroups of the appropriate great group of Chernozemic soils have one or more of the following characteristics: gley features within 50 cm, although the soils fail to meet criteria of the Gleysolic order; and low chromas or mottles (or both) below a depth of 50 cm.

Cryosolic Some Cryosolic soils have matrix colors of low chroma and prominent mottling within 50 cm of the surface similar to Gleysolic soils. Gleysolic soils, however, do not have permafrost within 1 m of the surface or 2 m if the soil is strongly cryoturbated.

Luvisolic Some soils have eluvial horizons, Bt horizons, and colors that indicate gleying within 50 cm of the mineral surface. Such soils are classified as Luvic Gleysols if gley colors as specified for the Gleysolic order occur in the Btg horizon within 50 cm of the mineral soil surface. If such gley colors occur only in the Aeg horizon (with the exception of thick Ae horizons as specified) or only below a depth of 50 cm, the soil is classified as a Gleyed subgroup of the appropriate great group in the Luvisolic order.

Organic Gleysolic soils may have organic surface layers, but they are too thin to meet the minimum limits specified for soils of the Organic order.

Podzolic A podzolic B horizon takes precedence over gley features. Thus, soils having both a podzolic B horizon and evidence of gleying that satisfies the specifications of Gleysolic soils are classified as Podzolic.

Regosolic Soils with no horizon differentiation, apart from evidence of gleying as specified for Gleysolic soils, are classified as Gleysolic.

Solonetzic Soils with both a Bn or Bnt horizon and evidence of gleying as specified for Gleysolic soils are classified as Solonetzic subgroups of the appropriate great groups of the Gleysolic order.

Vertisolic Some Gleysolic soils have a slickenside horizon but Vertisolic soils have both a vertic and a slickenside horizon. Gleysolic soils do not.

Gleysolic soils are divided into three great groups: Luvic Gleysol, Humic Gleysol, and Gleysol, which are separated based on the development of the Ah horizon and the presence or absence of a Bt horizon as shown in the Gleysolic order chart.

	Luvic Gleysol	Humic Gleysol	Gleysol
A horizon	usually an Ahe or an Aeg horizon	Ah horizon at least 10 cm thick	no Ah horizon or an Ah horizon <10 cm thick
B horizon	Btg horizon	no Bt horizon	no Bt horizon

Luvic Gleysol

Soils of this great group have the general properties specified for the Gleysolic order and a horizon of clay accumulation (Btg). They are similar to Luvisolic soils except that they have dull colors or prominent mottling, or both, which indicates strong gleying. They may have organic surface horizons and an Ah horizon. Luvic Gleysols occur commonly on poorly drained sites in association with Luvisolic soils and in depressions in areas of Black and Dark Gray Chernozemic soils.

Luvic Gleysols usually have an eluvial horizon (Ahe, Aeg) and a Btg horizon. A Btg horizon is defined based on an increase in silicate clay over that present in the A horizon. The presence of clay skins indicates illuvial clay. Colors and mottling as specified for the Gleysolic order indicate permanent or periodic reduction. Luvic Gleysols may have an organic surface horizon and an Ah horizon. In some cases the A horizon is very dark (value of 2) when moist, but its eluvial features are usually evident on drying. Such horizons generally have darker and lighter gray streaks and splotches, which are similar to the Ahe horizons of Dark Gray Chernozemic soils. Even if the eluvial horizon is dark, the Btg horizon is diagnostic of a Luvic Gleysol.

The great group is divided into six subgroups based on the kind and sequences of the horizons.

Vertic Luvic Gleysol

Common horizon sequence: LFH or O, Ah, Aeg, Bgf, Btg or Btgvj, Bntg or Bntgvj, Btgss or Cgss, Cg

These soils have the general properties specified for the Gleysolic order and the Luvic Gleysol great group. They may have any horizons in the upper solum common to other subgroups within the Luvic Gleysol great group. However, they differ from other subgroups by having properties indicative of intergrading to the Vertisolic order.

Specifically, they have a slickenside horizon (Btgss or Cgss), the upper boundary of which occurs within 1 m of the mineral soil surface. They also may have a weak vertic horizon (Btgvj or Bntgvj).

Solonetzic Luvic Gleysol

Common horizon sequence: LFH or O, Ah, Aeg, Bntg, Cg

These soils have the general properties specified for the Gleysolic order and the Luvic Gleysol great group. They have, in addition, a solonetzic B horizon. They may have Ah or Ap horizons as specified for Humic Luvic Gleysols. These soils are commonly associated with saline parent materials.

Fragic Luvic Gleysol

Common horizon sequence: LFH or O, Ah, Aeg, Btgx or Btg and Bxg, Cg

These soils have the general properties specified for the Gleysolic order and the Luvic Gleysol great group. They have, in addition, a fragipan within or below the Btg horizon. They may also have a dark-colored Ah or Ap horizon, as specified for Humic Luvic Gleysols, or a Bgf or Btgf horizon, as specified for Fera Luvic Gleysols. They do not have a solonetzic B horizon; such horizons are not known to occur in association with a fragipan.

Humic Luvic Gleysol

Common horizon sequence: LFH or O, Ah, Aeg, Btg, Cg

These soils have the general properties specified for the Gleysolic order and the Luvic Gleysol great group. They have, in addition, a mineral–organic surface horizon that meets the requirements of the Ah or Ap horizon of Humic Gleysols. Thus, the Ah horizon must be at least 10 cm thick and the Ap horizon must be at least 15 cm thick, contain at least 2% organic carbon, and have a darker color than the underlying horizon. Humic Luvic Gleysols have neither a solonetzic B horizon nor a fragipan, but they may have a Bgf horizon.

Fera Luvic Gleysol

Common horizon sequence: LFH or O, Ah, Aeg, Btgf or Bfg and Btg, Cg

These soils have the general properties specified for the Gleysolic order and the Luvic Gleysol great group. They also have either a Bgf horizon at least 10 cm thick as well as a Btg horizon or a Btgf horizon. A Bgf or Btgf horizon contains accumulated hydrous iron oxide (dithionite extractable), which is thought to have been deposited as a result of the oxidation of ferrous iron. It usually has a high chroma and is commonly a horizon with a concentration of rusty mottles. Fera Luvic Gleysols lack all the following: an Ah or Ap horizon diagnostic of Humic Luvic Gleysols, a solonetzic B horizon, and a fragipan.

Orthic Luvic Gleysol

Common horizon sequence: LFH or O, Aeg, Btg, Cg

These soils have the general properties specified for the Gleysolic order and the Luvic Gleysol great group. Typically, they have a Btg horizon and organic or mineral–organic surface horizons overlying gleyed, eluvial horizons.

Orthic Luvic Gleysols are identified by the following properties:

1. They have an eluvial horizon: Ahe, Ae, Aeg.
2. They have a Btg horizon.
3. They do not have an Ah or Ap horizon as defined for Humic Gleysols and Humic Luvic Gleysols.
4. They have neither a solonetzic B horizon, nor a fragipan, nor a Bgf horizon at least 10 cm thick.

Humic Gleysol

Soils of this great group have a dark-colored A horizon in addition to the general properties of soils of the Gleysolic order. They occur commonly in poorly drained positions in association with some Chernozemic, Luvisolic, Podzolic, and Brunisolic soils.

These soils may have organic surface horizons derived from grass and sedge, moss, or forest vegetation.

Humic Gleysols have no Bt horizon. They have either an Ah horizon at least 10 cm thick or a mixed surface horizon (Ap) at least 15 cm thick with all the following properties:

1. It contains at least 2% organic carbon.
2. It has a rubbed color value of 3.5 or less (moist), or 5.0 or less (dry).
3. It has a color value (moist) at least 1.5 units lower than that of the next underlying horizon if the color value (moist) of that horizon is 4 or more, or one unit of color value lower than that of the underlying horizon if its color value is less than 4.

Examples of color values of cultivated Humic Gleysols are as follows:

	Example 1	Example 2
Moist color value of Ap	≤ 3.5	≤ 2.0
Moist color value of underlying horizon	≥ 5.0	≥ 3.0

The great group is divided into five subgroups based on the kind of horizons and their sequence.

Vertic Humic Gleysol

Common horizon sequence: LFH or O, Ah, Bg or Bgvj, Bgss or Cgss, Cg

These soils have the general properties specified for the Gleysolic order and the Humic Gleysol great group. They may have any horizons in the upper solum common to other subgroups within the Humic Gleysol great group. However, they differ from other subgroups by having properties indicative of intergrading to the Vertisolic order. Specifically, they have a slickenside horizon (Bgss or Cgss), the upper boundary of which occurs within 1 m of the mineral soil surface. They may also have a weak vertic horizon (Bgvj).

Solonetzic Humic Gleysol

Common horizon sequence: Ah, Bng, Cgsk

These soils have the general properties specified for the Gleysolic order and the Humic Gleysol great group. They also have a solonetzic B horizon and may have a Bgf horizon. Typically, they have saline parent materials.

Fera Humic Gleysol

Common horizon sequence: LFH or O, _Ah,_ Aeg, _Bgf,_ Cg

These soils have the general properties specified for the Gleysolic order and the Humic Gleysol great group. They also have a Bgf horizon at least 10 cm thick and lack a solonetzic B horizon. The Bgf horizon contains accumulated hydrous iron oxide (dithionite extractable) thought to have been deposited as a result of the oxidation of ferrous iron. The Bgf horizon usually has many prominent mottles of high chroma.

Orthic Humic Gleysol

Common horizon sequence: LFH or O, _Ah,_ _Bg,_ Cg

These soils have the general properties specified for the Gleysolic order and the Humic Gleysol great group. Typically, they have a well-developed Ah horizon overlying gleyed B and C horizons. They may have organic surface horizons, an eluvial horizon, and a C horizon that does not have the dull colors and mottling that indicate gleying.

Orthic Humic Gleysols are identified by the following properties:

1. They have an Ah horizon at least 10 cm thick as defined for the great group.
2. They have a B horizon (Bg or Bgtj) at least 10 cm thick.
3. They do not have any of the following: a Btg horizon, a solonetzic B horizon, or a Bgf horizon at least 10 cm thick.

Rego Humic Gleysol

Common horizon sequence: LFH or O, _Ah,_ _Cg_

These soils have the general properties specified for the Gleysolic order and the Humic Gleysol great group. They differ from Orthic Humic Gleysols in that they lack a B horizon at least 10 cm thick. Typically, they have a well-developed Ah horizon overlying a gleyed C horizon.

Gleysol

Soils of this great group have the general properties specified for soils of the Gleysolic order, but they lack a well-developed, mineral–organic surface horizon. They occur commonly in poorly drained positions in association with soils of several other orders.

Gleysols lack an Ah or Ap horizon as specified for Humic Gleysols. They also lack a Bt horizon. They may have either an Ah horizon thinner than 10 cm or an Ap horizon with one of the following properties:

1. It contains less than 2% organic carbon.
2. It has a rubbed color value greater than 3.5 (moist) or greater than 5.0 (dry).
3. It shows little contrast in color value with the underlying layer (a difference of less than 1.5 units if the value of the underlying layer is 4 or more, or a difference of less than 1 unit if that value is less than 4).

These soils have a gleyed B or C horizon, and they may have an organic surface horizon.

The great group is divided into five subgroups based on the kind of horizons and their sequence.

Vertic Gleysol

Common horizon sequence: LFH or O, _Bg_ or _Bgvj,_ _Bgss_ or _Cgss,_ Cg

These soils have the general properties specified for the Gleysolic order and the Gleysol great group. They may have any horizons in the upper solum common to other subgroups within the Gleysol great group. However, they differ from other subgroups by having properties indicative of intergrading to the Vertisolic order. Specifically, they have a slickenside horizon (Bgss or Cgss), the upper boundary of which occurs within 1 m of the mineral soil surface. They may also have a weak vertic horizon (Bgvj).

Solonetzic Gleysol

Common horizon sequence: LFH or O, _Aeg,_ _Bng,_ Cgsk

These soils have the general properties specified for the Gleysolic order and the Gleysol great group. They also have a solonetzic B horizon and may have a Bgf horizon. Typically, they have saline parent materials.

Fera Gleysol

Common horizon sequence: LFH or O, Aeg, <u>Bgf</u>, Cg

These soils have the general properties specified for the Gleysolic order and the Gleysol great group. They also have a Bgf horizon at least 10 cm thick and lack a solonetzic B horizon. The Bgf horizon contains accumulated hydrous iron oxide (dithionite extractable), which is thought to have been deposited as a result of the oxidation of ferrous iron. The Bgf horizon usually has many prominent mottles of high chroma.

Orthic Gleysol

Common horizon sequence: LFH or O, <u>Bg</u>, Cg

These soils have the general properties specified for the Gleysolic order and the Gleysol great group. Typically, they have strongly gleyed B and C horizons and may have organic surface horizons and an eluvial horizon. Orthic Gleysols are identified by the following properties:

1. They have a B horizon (Bg or Btjg) at least 10 cm thick.

2. They may have an Ah or Ap horizon as specified for the Gleysol great group.

3. They have neither a Btg horizon, nor a solonetzic B horizon, nor a Bgf horizon at least 10 cm thick.

Rego Gleysol

Common horizon sequence: LFH or O, <u>Cg</u>

These soils have the general properties specified for the Gleysolic order and the Gleysol great group. They differ from Orthic Gleysols in that they lack a B horizon at least 10 cm thick. They thus consist of a gleyed C horizon, with or without organic surface horizons, and a thin Ah or B horizon.

Chapter 8

Luvisolic Order

Great Group	Subgroup
Gray Brown Luvisol	Orthic Gray Brown Luvisol O.GBL
	Brunisolic Gray Brown Luvisol BR.GBL
	Podzolic Gray Brown Luvisol PZ.GBL
	Vertic Gray Brown Luvisol V.GBL
	Gleyed Gray Brown Luvisol GL.GBL
	Gleyed Brunisolic Gray Brown Luvisol GLBR.GBL
	Gleyed Podzolic Gray Brown Luvisol GLPZ.GBL
	Gleyed Vertic Gray Brown Luvisol GLV.GBL
Gray Luvisol	Orthic Gray Luvisol O.GL
	Dark Gray Luvisol D.GL
	Brunisolic Gray Luvisol BR.GL
	Podzolic Gray Luvisol PZ.GL
	Solonetzic Gray Luvisol SZ.GL
	Fragic Gray Luvisol FR.GL
	Vertic Gray Luvisol V.GL
	Gleyed Gray Luvisol GL.GL
	Gleyed Dark Gray Luvisol GLD.GL
	Gleyed Brunisolic Gray Luvisol GLBR.GL
	Gleyed Podzolic Gray Luvisol GLPZ.GL
	Gleyed Solonetzic Gray Luvisol GLSZ.GL
	Gleyed Fragic Gray Luvisol GLFR.GL
	Gleyed Vertic Gray Luvisol GLV.GL

A diagrammatic representation of profiles of some subgroups of the Luvisolic order is shown in Figure 33. Individual subgroups may include soils that have horizon sequences different from those shown. In the description of each subgroup, presented later in this chapter, a common horizon sequence is given; diagnostic horizons are underlined and some other commonly occurring horizons are listed.

Soils of the Luvisolic order generally have light-colored, eluvial horizons and have illuvial B horizons in which silicate clay has accumulated. These soils develop characteristically in well to imperfectly drained sites, in sandy loam to clay, base-saturated parent materials under forest vegetation in subhumid to humid, mild to very cold climates. Depending on the combination of soil environmental factors, some Luvisolic soils occur under conditions outside the range indicated as characteristic. For example, some Luvisolic soils develop in acid parent materials and some occur in forest–grassland transition zones.

Luvisolic soils occur from the southern extremity of Ontario to the zone of permafrost and from Newfoundland to British Columbia. The largest area of these soils occurs in the central to northern Interior Plains under deciduous, mixed, and coniferous forest. The Gray Luvisols of that area usually have well-developed, platy Ae horizons of low chroma, Bt horizons with moderate to strong prismatic or blocky structures, calcareous C horizons, and sola of high base saturation (neutral salt extraction). Gray Luvisols of the Atlantic Provinces commonly have Bt horizons of weak structure and low to moderate base saturation. The Gray Brown Luvisols of southern Ontario and some valleys of British Columbia characteristically have forest-mull Ah horizons, moderate to strong blocky structured Bt horizons, and calcareous C horizons.

Luvisolic soils have an eluvial and Bt horizon as defined. The Bt horizon must have the specified increase in clay over that in the

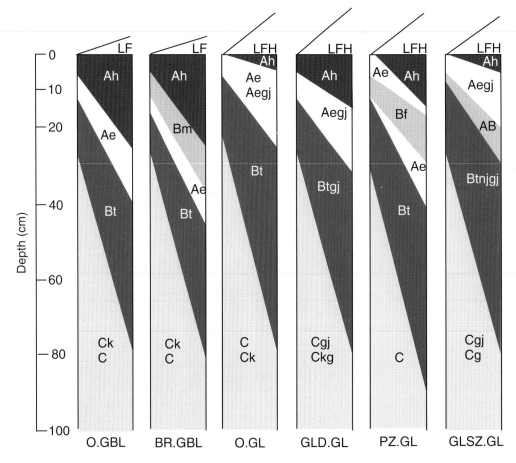

Figure 33 Diagrammatic horizon pattern of some subgroups of the Luvisolic order.

eluvial horizon, clay skins indicative of translocated clay accounting for 1% or more of the area of a section through the horizon, and be at least 5 cm thick. Luvisolic soils may have Ah, Ahe, or dark-colored Ap horizons that satisfy one or more of the following conditions:

1. The dark-colored A horizon does not meet the requirements of a chernozemic A.

2. The dark-colored A horizon is underlain by a thicker, light-colored Ae horizon that extends to a depth of 15 cm from the mineral surface.

3. The dark-colored A horizon shows evidence of eluviation (Ahe or Ap) and is underlain by an Ae horizon at least 5 cm thick.

4. If the soil moisture regime is humid or wetter, the dark-colored A horizon may be of any kind.

Luvisolic soils have none of the following: a solonetzic B horizon, a podzolic B horizon if the upper boundary of the Bt horizon is at a depth of more than 50 cm from the mineral surface, evidence of gleying strong enough to meet the requirements of the Gleysolic order, organic horizons thick enough to meet the requirements of the Organic order, or permafrost within 1 m of the mineral surface and 2 m if the soils are cryoturbated.

The genesis of Luvisolic soils is thought to involve the suspension of clay in the soil solution near the soil surface, downward movement of that clay with the soil solution, and the deposition of the translocated clay at a depth where downward movement of the soil solution ceases or becomes very slow. Commonly, the subhorizon of maximum clay accumulation occurs above a Ck horizon. The eluvial horizon (Ahe, Ae) commonly has platy structure due perhaps to the periodic formation of ice lenses. Any condition that promotes dispersion of clay in the A horizons and deposition of this clay in a discrete subsurface horizon favors the development of Luvisolic soils.

Distinguishing Luvisolic Soils from Soils of Other Orders

Guidelines for distinguishing Luvisolic soils from soils of other orders with which they might be confused follow.

Chernozemic Some Chernozemic soils have Ah, Ae, and Bt horizons as do some Luvisolic soils. A basis of differentiation of these soils is the nature of the Ah and Ae horizons as follows:

1. If the Ah is not a chernozemic A, the soil is Luvisolic.
2. If the soil has a chernozemic A and a light-colored Ae that is thicker than the Ah or Ap and extends to a depth below 15 cm, the soil is Luvisolic.
3. If the soil has an eluviated, dark-colored A horizon (Ahe or Ap) and a distinct Ae horizon thicker than 5 cm or that extends below the Ap, it is Luvisolic.

Solonetzic Solonetzic soils have a solonetzic B horizon, but Luvisolic soils do not.

Podzolic Some soils have both a podzolic B and a Bt horizon. The soil is classified as Luvisolic if the upper boundary of the Bt horizon is within 50 cm of the mineral surface and as Podzolic if the boundary is more than 50 cm below the surface.

Brunisolic Luvisolic soils have a Bt horizon, but Brunisolic soils do not.

Gleysolic Some Gleysolic soils have Ae and Bt horizons, but unlike Luvisolic soils they also have colors of low chroma or prominent mottling indicative of strong gleying.

Vertisolic Some Luvisolic soils have a slickenside horizon but Vertisolic soils have both a vertic and a slickenside horizon. Luvisolic soils do not.

Luvisolic soils are divided into two great groups, Gray Brown Luvisol and Gray Luvisol as shown in the Luvisolic order chart. The latter accounts for most of the area of Luvisolic soils.

Subgroups are separated based on the kind and sequence of horizons indicating conformity with the central concept of the great group or intergrades to soils of other orders, or additional features. The former subgroup features turbic, cryic, and lithic are now recognized taxonomically at either the family or series level. They may be indicated also as phases of subgroups, great groups, or orders.

Gray Brown Luvisol

Soils of this great group have a forest mull Ah horizon and eluvial and Bt horizons as indicated for the order. They occur typically under deciduous or mixed forest vegetation on calcareous materials in areas of mild, humid climate. They occur mainly in the St. Lawrence Lowland.

Under undisturbed conditions the soils may have thin L, F, and H horizons, but, because of high biological activity and the abundance of earthworms, the leaf litter is usually quickly incorporated into the soil and humified. A transitional AB or BA horizon having gray coatings on structural aggregates is generally present, particularly in medium- and fine-textured soils. Although the Bt horizon is generally immediately underlain by calcareous materials, a transitional BC horizon may be present.

Under cultivated conditions the Ah and commonly part of the Ae horizon are mixed to form an Ap horizon. However the Bt and part of the Ae horizon usually remain intact under the Ap horizon unless either cultivation greatly exceeds 15 cm in depth or surface erosion has caused the Ap horizon to be developed in the Ae or Bt horizons or both.

Gray Brown Luvisols have either a forest-mull Ah horizon more than 5 cm thick or a dark-colored Ap horizon, an eluvial horizon, and a Bt horizon. The mean annual soil temperature is 8°C or higher and the soil moisture regime is humid or wetter.

Luvisolic Order

	Gray Brown Luvisol	Gray Luvisol
A horizon	forest-mull Ah	may or may not have Ah
B horizon	eluvial and Bt horizons	eluvial and Bt horizons
Mean annual soil temperature	≥8°C	usually <8°C

The great group is divided into eight subgroups based on the profile developed above the Bt horizon, the presence of Vertic features, and evidence of gleying.

Orthic Gray Brown Luvisol

Common horizon sequence: Ah, Ae, Bt, Ck

These soils have the properties specified for the Luvisolic order and the Gray Brown Luvisol great group. They have well-developed Ah, eluvial, and Bt horizons, and usually calcareous C horizons. Faint mottling may occur immediately above or within the Bt horizon.

Orthic Gray Brown Luvisols are identified by the following properties:

1. These soils have either a forest-mull Ah horizon more than 5 cm thick or a dark-colored (moist) Ap horizon.

2. These soils have an Ae horizon of which the upper 5 cm is light colored with a chroma of 3 or less. The difference in chroma between the upper and lower part of the Ae is less than 1.

3. These soils have a Bt horizon and lack a Bf horizon.

4. Distinct mottling indicative of gleying does not occur within 50 cm of the mineral surface, and prominent mottling does not occur at depths of 50–100 cm.

Brunisolic Gray Brown Luvisol

Common horizon sequence: Ah, Ae, Bm or Bf, Ae, Bt, BC, Ck

These soils have the properties specified for the Luvisolic order and the Gray Brown Luvisol great group. They differ from Orthic Gray Brown Luvisols by having in the upper solum either a Bm horizon at least 5 cm thick with a chroma of 3 or more, or a Bf horizon less than 10 cm thick that does not extend below 15 cm. Such Bm or Bf horizons are thought to have developed in a former Ae horizon. If disturbance results in the Bm or Bf horizon being incorporated into the Ap, the disturbed soil is classified as an Orthic Gray Brown Luvisol.

Podzolic Gray Brown Luvisol

Common horizon sequence: LFH, Ah, Ae, Bf, Ae, Bt, BC, Ck

These soils have the properties specified for the Luvisolic order and the Gray Brown Luvisol great group. They differ from Orthic Gray Brown Luvisols by having a Bf horizon at least 10 cm thick in the upper solum. These soils may or may not have an Ae horizon. The upper boundary of the Bt horizon must be within 50 cm of the mineral surface or the soil is classified in the Podzolic order. If disturbance results in the Bf horizon being incorporated into the Ap, the disturbed soil is classified as an Orthic Gray Brown Luvisol.

Vertic Gray Brown Luvisol

Common horizon sequence: Ah, Ae, Bm or Bf, Bt, Bvj, Bss or Ckss, Ck

These soils have the general properties specified for the Luvisolic order and the Gray Brown Luvisol great group. They may have any horizons in the upper solum common to other subgroups within the Gray Brown Luvisol great group. However, they differ from other subgroups by having properties indicative of intergrading to the Vertisolic order. Specifically, they have a slickenside horizon (Bss, Bkss, or Ckss), the upper boundary of which occurs within 1 m of the mineral surface. They may have a weak vertic horizon (Bvj).

Gleyed Gray Brown Luvisol

Common horizon sequence: Ah, Aegj, Btgj, Ckg

These soils have the properties specified for the Luvisolic order and the Gray Brown Luvisol great group. They differ from Orthic Gray Brown Luvisols by having either distinct mottles that indicate gleying within 50 cm of the mineral surface, or prominent mottles at depths of 50–100 cm. Commonly the matrix colors are of lower chroma than those of associated, well-drained soils. Gleyed Gray Brown Luvisols generally have thicker and darker Ah horizons than well-drained subgroups on similar parent materials. The color and textural differences between the Ae and Bt horizons are generally less marked in the gleyed than in the orthic subgroup.

Gleyed Brunisolic Gray Brown Luvisol

Common horizon sequence: <u>Ah</u>, <u>Bmgj</u>, Aegj, <u>Btgj</u>, Ckg

These soils have the properties specified for the Luvisolic order and the Gray Brown Luvisol great group. They differ from Brunisolic Gray Brown Luvisols by having either distinct mottles that indicate gleying within 50 cm of the mineral surface, or prominent mottles at depths of 50–100 cm.

Gleyed Podzolic Gray Brown Luvisol

Common horizon sequence: <u>Ah</u>, Ae, <u>Bfgj</u>, Aegj, <u>Btgj</u>, Ckg

These soils have the properties specified for the Luvisolic order and the Gray Brown Luvisol great group. They differ from Podzolic Gray Brown Luvisols by having either distinct mottles that indicate gleying within 50 cm of the mineral surface, or prominent mottles at depths of 50–100 cm. Besides a forest-mull Ah horizon, they have a Bf horizon at least 10 cm thick and a gleyed Bt horizon of which the upper boundary is within 50 cm of the mineral surface.

Gleyed Vertic Gray Brown Luvisol

Common horizon sequence: <u>Ah</u>, Ae, <u>Bmgj</u>, or <u>Bfgj</u>, <u>Btgj</u>, Bgjvj, <u>Bgjss</u> or <u>Ckgjss</u>, Ckgj or Ckg

These soils have the general properties specified for the Luvisolic order and the Gray Brown Luvisol great group. Also, they have properties that indicate intergrading to the Vertisolic order. Specifically, they have a slickenside horizon (Bgjss or Ckgjss), the upper boundary of which occurs within 1 m of the mineral surface. These soils may have a weak vertic horizon (Bgjvj). They differ from Vertic Gray Brown Luvisols by having either distinct mottles that indicate gleying within 50 cm of the mineral surface, or prominent mottles at depths of 50–100 cm.

Gray Luvisol

Soils of this great group have eluvial and Bt horizons as specified for the Luvisolic order. They usually have L, F, and H horizons and may have a degraded Ah or Ahe horizon that resembles the upper A horizon of Dark Gray Chernozemic soils. Commonly below the Ae horizon they have an AB or BA horizon in which the ped surfaces are grayer than the interiors of peds. The solum of Gray Luvisols is generally slightly to moderately acid but may be strongly acid. The degree of base saturation (neutral salt extraction) is generally high. The parent materials are usually base saturated and commonly calcareous, but some Gray Luvisols have developed in acid materials.

Gray Luvisols occur typically under boreal or mixed forest vegetation and in forest–grassland transition zones in a wide range of climatic areas. Their main area of occurrence is in the subhumid central to northern part of the Interior Plains, but they occur also in humid and perhumid areas of eastern Canada. In the latter climatic area, they occur mainly on medium- to fine-textured parent materials.

Gray Luvisols have eluvial and Bt horizons. Their mean annual soil temperature is usually less than 8°C. If the soil moisture regime is subhumid, any dark-colored A horizon must satisfy one or more of the following conditions:

1. It is not a chernozemic A.

2. It is underlain by a thicker Ae horizon that extends to a depth greater than 15 cm below the mineral surface.

3. It shows evidence of degradation (Ahe) and is underlain by an Ae horizon at least 5 cm thick below the Ahe or Ap horizon.

The great group is divided into 14 subgroups based on the kind and sequence of horizons and evidence of gleying.

Orthic Gray Luvisol

Common horizon sequence: LFH, <u>Ae</u>, AB, <u>Bt</u>, C or Ck

These soils have the properties specified for the Luvisolic order and the Gray Luvisol great group. They have well-developed Ae and Bt horizons and usually have organic surface horizons. Faint mottling may occur immediately above or within the Bt horizon.

Orthic Gray Luvisols are identified by the following properties:

1. They have an Ae horizon with a chroma of less than 3 unless the chroma of the parent material is 4 or more.

2. They have a Bt horizon.

3. They lack a Bf horizon.

4. They lack a fragipan.

5. They may have a dark-colored, mineral–organic surface horizon (Ah or Ahe) less than 5 cm thick.

6. They may have an Ap horizon, but its dry color value must be 5 or higher.

7. Distinct mottling, that indicates gleying does not occur within 50 cm of the mineral surface. Prominent mottling does not occur at depths of 50–100 cm.

Dark Gray Luvisol

Common horizon sequence: LFH, Ah or Ahe, Ae, Bt, C or Ck

These soils have the properties specified for the Luvisolic order and the Gray Luvisol great group. They differ from Orthic Gray Luvisols by having an Ah or Ahe horizon 5 cm or more in thickness. These horizons generally have eluvial features, such as gray streaks or splotches when dry, or platy structure. In the case of disturbed soils, the dry color value of the Ap horizon is 3.5–5.0 and some of the Ae horizon remains below the Ap. The differentiation of Dark Gray Luvisols from Dark Gray Chernozemic soils was outlined previously in this chapter. Dark Gray Luvisols have a mean annual soil temperature less than 8°C. They may have a Btnj or a Bm horizon above the Bt horizon.

Brunisolic Gray Luvisol

Common horizon sequence: LFH, Bm or Bf, Ae, Bt, BC, C or Ck

These soils have the properties specified for the Luvisolic order and the Gray Luvisol great group. They differ from Orthic Gray Luvisols by having in the upper solum, either a Bm horizon at least 5 cm thick with a chroma of 3 or more, or a Bf horizon less than 10 cm thick that does not extend below 15 cm. Such Bm or Bf horizons are thought to have developed in a former Ae horizon. These soils do not have a dark-colored Ah or Ahe horizon 5 cm or more in thickness.

Podzolic Gray Luvisol

Common horizon sequence: LFH, Ae, Bf, Ae, Bt, BC, C or Ck

These soils have the properties specified for the Luvisolic order and the Gray Luvisol great group. They differ from Orthic Gray Luvisols by having a Bf horizon at least 10 cm thick in the upper solum. They may also have a dark-colored Ah or Ahe horizon 5 cm or more in thickness. The upper boundary of the Bt horizon must be within 50 cm of the mineral surface or the soil is classified in the Podzolic order.

Solonetzic Gray Luvisol

Common horizon sequence: LFH, Ae, AB, Btnj, BC, C or Csk

These soils have the properties specified for the Luvisolic order and the Gray Luvisol great group. They differ from Orthic Gray Luvisols by having a Btnj horizon that indicates an intergrade to the Solonetzic order. The Btnj horizon usually has a harder consistence, more pronounced coatings on the prismatic or blocky peds, and a lower ratio of exchangeable Ca:Na than the Bt horizons of most Gray Luvisols. This subgroup is associated with saline parent materials. Solonetzic Gray Luvisols do not have an Ah or Ahe horizon 5 cm or more in thickness, and so the surface horizons do not meet the criteria for Dark Gray Luvisols.

Fragic Gray Luvisol

Common horizon sequence: LFH, Ahe, Ae, Bt, Btx or BCx, C

These soils have the properties specified for the Luvisolic order and the Gray Luvisol great group. They differ from Orthic Gray Luvisols by having a fragipan either within or below the Bt horizon. They also differ form Orthic Gray Luvisols by having a dark-coloured Ah or Ahe horizon 5 cm or more in thickness and a Btnj, Bm, or Bf horizon.

Vertic Gray Luvisol

Common horizon sequence: LFH, Ah or Ahe, Ae, AB, Bt or Btnj, Btvj, Btss or Ckss, Ck

These soils have the general properties specified for the Luvisolic order and the Gray Luvisol great group. They may have any horizons in the upper solum common to other subgroups within the Gray Luvisol great group. However, they differ from other subgroups by having properties that indicate intergrading to the Vertisolic order. Specifically, they have a slickenside horizon (Btss or Ckss), the upper boundary of which occurs within 1 m of the mineral surface. They may have a weak vertic horizon (Btvj).

Gleyed Gray Luvisol

Common horizon sequence: LFH, <u>Ae</u>, <u>Btg</u>, Cg

These soils have the properties specified for the Luvisolic order and the Gray Luvisol great group. They differ from Orthic Gray Luvisols by having either distinct mottles that indicate gleying within 50 cm of the mineral surface, or prominent mottles at depths of 50–100 cm. Commonly the matrix colors are of lower chroma than those of associated, well-drained soils.

Gleyed Dark Gray Luvisol

Common horizon sequence: LFH, <u>Ah</u> or <u>Ahe</u>, <u>Ae</u>, <u>Btgj</u>, Cg or Ckg

These soils have the properties specified for the Luvisolic order and the Gray Luvisol great group. They differ from Dark Gray Luvisols by having either distinct mottles that indicate gleying within 50 cm of the mineral surface, or prominent mottles at depths of 50–100 cm.

Gleyed Brunisolic Gray Luvisol

Common horizon sequence: LFH, <u>Bm</u> or <u>Bf</u>, Aegj, <u>Btgj</u>, BCgj, Cg

These soils have the properties specified for the Luvisolic order and the Gray Luvisol great group. They differ from Brunisolic Gray Luvisols by having either distinct mottles that indicate gleying within 50 cm of the mineral surface, or prominent mottles at depths of 50–100 cm. They do not have Ah or Ahe horizons 5 cm or more in thickness.

Gleyed Podzolic Gray Luvisol

Common horizon sequence: LFH, Ae, <u>Bf</u>, Aegj, <u>Btgj</u>, BCgj, Cg

These soils have the properties specified for the Luvisolic order and the Gray Luvisol great group. They differ from Podzolic Gray Luvisols by having either distinct mottles that indicate gleying within 50 cm of the mineral surface, or prominent mottles at depths of 50–100 cm. They may have a dark-colored A horizon 5 cm or more in thickness.

Gleyed Solonetzic Gray Luvisol

Common horizon sequence: LFH, Ae, ABgj, <u>Btnjgj</u>, Cgj or Csag

These soils have the properties specified for the Luvisolic order and the Gray Luvisol great group. They differ from Solonetzic Gray Luvisols by having either distinct mottles that indicate gleying within 50 cm of the mineral surface, or prominent mottles at depths of 50–100 cm. They do not have an Ah or Ahe horizon 5 cm or more in thickness.

Gleyed Fragic Gray Luvisol

Common horizon sequence: LFH, Ahe, <u>Aegj</u>, <u>Btgj</u>, <u>Btxgj</u> or <u>BCxg</u>, Cg

These soils have the properties specified for the Luvisolic order and the Gray Luvisol great group. They differ from Fragic Gray Luvisols by having either distinct mottles that indicate gleying within 50 cm of the mineral surface, or prominent mottles at depths of 50–100 cm. They may have a dark-colored A horizon 5 cm or more in thickness and a Bm, Btnj, or Bf horizon.

Gleyed Vertic Gray Luvisol

Common horizon sequence: LFH, Ah or Ahe, <u>Ae</u>, AB, <u>Btgj</u> or <u>Btnjgj</u>, Btgjvj, <u>Btgjss</u> or <u>Ckgjss</u>, Ckgj or Ckg

These soils have the general properties specified for the Luvisolic order and the Gray Luvisol great group. They also have properties that indicate intergrading to the Vertisolic order. Specifically, they have a slickenside horizon (Btgjss or Ckgjss), the upper boundary of which occurs within 1 m of the mineral surface. These soils may have a weak vertic horizon (Btgjvj). They differ from Vertic Gray Luvisols by having either distinct mottles that indicate gleying within 50 cm of the mineral surface, or prominent mottles at depths of 50–100 cm.

Chapter 9

Organic Order

Great Group	Subgroup
Fibrisol	Typic Fibrisol TY.F
	Mesic Fibrisol ME.F
	Humic Fibrisol HU.F
	Limnic Fibrisol LM.F
	Cumulic Fibrisol CU.F
	Terric Fibrisol T.F
	Terric Mesic Fibrisol TME.F
	Terric Humic Fibrisol THU.F
	Hydric Fibrisol HY.F
Mesisol	Typic Mesisol TY.M
	Fibric Mesisol FI.M
	Humic Mesisol HU.M
	Limnic Mesisol LM.M
	Cumulic Mesisol CU.M
	Terric Mesisol T.M
	Terric Fibric Mesisol TFI.M
	Terric Humic Mesisol THU.M
	Hydric Mesisol HY.M
Humisol	Typic Humisol TY.H
	Fibric Humisol FI.H
	Mesic Humisol ME.H
	Limnic Humisol LM.H
	Cumulic Humisol CU.H
	Terric Humisol T.H
	Terric Fibric Humisol TFI.H
	Terric Mesic Humisol TME.H
	Hydric Humisol HY.H
Folisol	Hemic Folisol HE.FO
	Humic Folisol HU.FO
	Lignic Folisol LI.FO
	Histic Folisol HI.FO

A diagrammatic representation of the depth relationships of tiers, and of Typic and Terric subgroups of Organic soils, is shown in Figure 34. Diagrammatic sketches of profiles of some subgroups of the Organic order are shown in Figures 35 and 36. Individual subgroups may include soils that have horizon sequences different from those shown. In the description of each subgroup, presented later in this chapter, a common horizon sequence is given; diagnostic horizons or layers are underlined and some other commonly occurring horizons are listed.

Soils of the Organic order are composed largely of organic materials. They include most of the soils commonly known as peat, muck, or bog and fen soils. Most Organic soils are saturated with water for prolonged periods. These soils occur widely in poorly and very poorly drained depressions and level areas in regions of subhumid to perhumid climate and are derived from vegetation that grows in such sites. However, one group of Organic soils (Folisols) consists of upland (folic) organic materials, generally of forest origin. These Folisols are well to imperfectly drained, although they may become saturated after rainfall or snowmelt.

Organic soils contain more than 17% organic C (30% or more organic matter) by weight and meet the following specifications.

For organic materials (O) that are commonly saturated with water and consist mainly of mosses, sedges, or other hydrophytic vegetation the specifications are as follows:

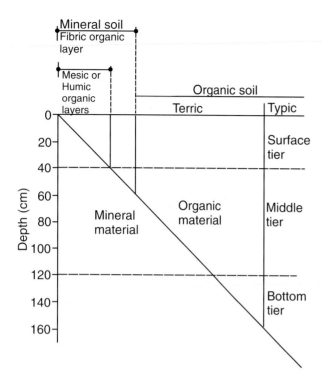

Figure 34 **Diagrammatic representation of depth relationships in the control section used to classify Fibrisol, Mesisol, and Humisol great groups.**

1. If the surface layer consists of fibric organic material with or without mesic or humic Op horizons thinner than 15 cm, the organic material must extend to a depth of at least 60 cm.

2. If the surface layer is mesic or humic, the organic material must extend to a depth of at least 40 cm.

3. If a lithic contact occurs at a depth shallower than 40 cm, the organic material must extend to a depth of at least 10 cm. Mineral material less than 10 cm thick may overlie the lithic contact, but the organic material must be more than twice the thickness of the mineral layer.

4. The organic soil may have a mineral layer thinner than 40 cm on the surface provided that the underlying organic material is at least 40 cm thick.

5. Mineral layers thinner than 40 cm that begin within a depth of 40 cm from the surface may occur within an Organic soil. A mineral layer or layers with a combined thickness of less than 40 cm may occur within the upper 80 cm.

For folic materials (L, F, and H) not usually saturated with water there must be

1. Forty centimetres or more of folic materials if directly overlying mineral soil or peat materials.

2. Greater than 10 cm of folic materials if directly overlying a lithic contact or fragmental materials.

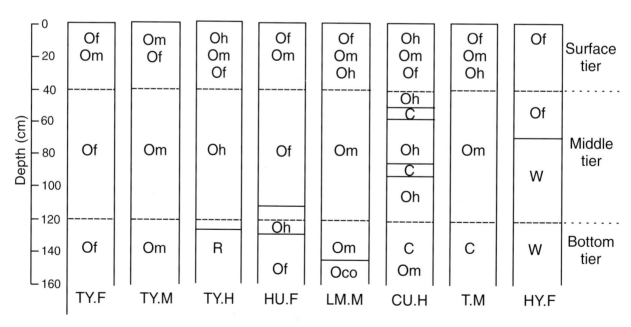

Figure 35 **Diagrammatic horizon pattern of some subgroups of the Fibrisol, Mesisol, and Humisol great groups.**

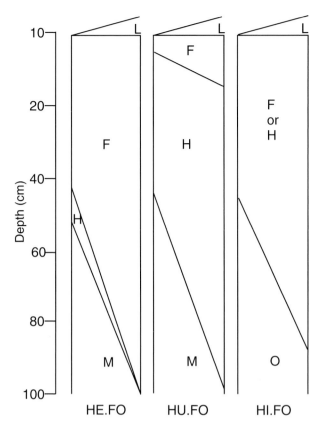

Figure 36 Diagrammatic horizon pattern of some subgroups of the Folisol great group.

3. More than twice the thickness of a mineral soil layer if the mineral layer is less than 20 cm thick.

The control section (160 cm) for Fibrisol, Mesisol, and Humisol great groups is divided into three tiers: surface (0–40 cm); middle (40–120 cm); and bottom (120–160 cm) (see Chapter 2 for detailed definitions). Classification at the great group level is based primarily on properties of the middle tier.

Distinguishing Organic Soils from Soils of Other Orders

Many soils of other orders may have organic horizons at the surface. The distinction

between Organic soils and soils of other orders is based on the following:

1. The thickness and the organic C content of organic-rich surface horizons in the case of soils with O horizons.

2. The thickness of the folic material for soils with L, F, and H horizons.

3. The depth to permafrost; organic materials having permafrost at depths of 1 m or less are classified as Cryosolic soils.

Organic soils are divided into four great groups as indicated in the Organic order chart. Three of these represent Organic soils formed in hydrophytic vegetation and are separated on the basis of degree of decomposition of the organic material. These soils are commonly saturated with water throughout the year. The fourth represents organic soils formed in upland (folic) organic materials and are soils that are only briefly saturated with water.

Subgroups are based upon the kinds and sequences of horizons.

Fibrisol

Soils of this great group are composed largely of relatively undecomposed fibric organic material. Fibric material is usually classified on the von Post scale of decomposition as classes 1–4. Fibrisols occur extensively in Canada, particularly in peat deposits dominated by sphagnum mosses.

Fibrisols have a dominantly fibric middle tier, or middle and surface tiers if a terric, lithic, or hydric contact occurs in the middle tier. Fibric material is the least decomposed type of organic material. It contains large amounts of well-preserved fiber that is retained on a 100-mesh sieve (0.15 mm) and can be identified as to botanical origin. A fibric horizon has 40% or more of rubbed fiber by volume and a pyrophosphate index of 5 or more (*see* Chapter 2, Organic horizons, Of). If the rubbed fiber volume is 75% or more, the pyrophosphate criterion does not apply. Dominantly fibric means that fibric material is

Organic Order			
Hydrophytic Vegetation			Upland Organic Material
Fibrisol	Mesisol	Humisol	Folisol
Fibric middle tier	Mesic middle tier	Humic middle tier	Folic materials, rarely saturated with water

the most abundant type of organic material. If both fibric and mesic layers occur in the middle tier, the tier is dominantly fibric if more than half of its thickness is composed of fibric material. If fibric, mesic, and humic layers are present in the middle tier, it is dominantly fibric if the thickness of fibric layers is greater than that of either mesic or humic layers. Subdominant in the following definitions means next in abundance to the dominant material but the layer must not be less than 12 cm in thickness if sharply contrasting (Of vs Oh), or 25 cm in thickness if not sharply contrasting (Om vs Of or Oh).

Typic Fibrisol

Common horizon sequence: Of or Om, <u>Of</u>

Soils of this subgroup have the general properties specified for the Organic order and the Fibrisol great group. They are composed mainly of fibric material that is commonly derived mainly from mosses.

These soils are identified by the following properties:

1. If present, the middle and bottom tiers are dominantly fibric. A lithic contact may occur.
2. They have neither subdominant humic layers with a total thickness of greater than 12 cm or subdominant mesic layers with a total thickness greater than 25 cm in the middle and bottom tier, or in the middle and surface tiers if a lithic contact occurs in the middle tier.
3. The middle tier lacks terric, hydric, cumulic, and limnic layers.

Mesic Fibrisol

Common horizon sequence: Of or Om, <u>Of</u>, <u>Om</u>, Of

Soils of this subgroup have the general properties specified for the Organic order and the Fibrisol great group. They differ from Typic Fibrisols by having a subdominant mesic layer (thicker than 25 cm) in the middle or bottom tier. The control section lacks terric, hydric, cumulic, or limnic layers.

Humic Fibrisol

Common horizon sequence: Of, Om or Oh, <u>Of</u>, <u>Oh</u>, Of or Om

Soils of this subgroup have the general properties specified for the Organic order and

the Fibrisol great group. They differ from Typic Fibrisols by having a subdominant humic layer thicker than 12 cm in the middle or bottom tier. They may also have a subdominant mesic layer. The control section lacks terric, hydric, cumulic, and limnic layers.

Limnic Fibrisol

Common horizon sequence: Of or Om, <u>Of</u>, <u>Oco</u>

Soils of this subgroup have the general properties specified for the Organic order and the Fibrisol great group. They differ from Typic Fibrisols by having a limnic layer beneath the surface tier. A limnic layer is a layer or layers at least 5 cm thick of coprogenous earth (sedimentary peat), diatomaceous earth, or marl. Limnic materials include both organic and inorganic materials either deposited in water by precipitation or by the action of aquatic organisms such as algae and including diatoms, or derived from underwater and floating aquatic plants subsequently modified by aquatic animals. Except for some of the coprogenous earths, most of these limnic materials are inorganic. Diatomaceous earth is highly siliceous and marl is mainly $CaCO_3$. Limnic Fibrisols may have mesic, humic, or cumulic layers but do not have terric or hydric layers.

Cumulic Fibrisol

Common horizon sequence: Of or Om, <u>Of</u>, <u>C</u>, Of

Soils of this subgroup have the general properties specified for the Organic order and the Fibrisol great group. They differ from Typic Fibrisols by having a cumulic layer beneath the surface tier. They may have mesic or humic layers but lack terric, hydric, and limnic layers. A cumulic layer consists either of multiple layers of mineral material (alluvium) that together are more than 5 cm thick, or of one layer 5–30 cm thick.

Terric Fibrisol

Common horizon sequence: Of or Om, <u>Of</u>, <u>C</u>

Soils of this subgroup have the general properties specified for the Organic order and the Fibrisol great group. They differ from Typic Fibrisols by having a terric layer (an unconsolidated mineral layer at least 30 cm thick) beneath the surface tier. They may also have cumulic or limnic layers but do not have mesic, humic, or hydric layers within the control section.

Terric Mesic Fibrisol

Common horizon sequence: Of or Om, <u>Of</u>, <u>Om</u>, C

Soils of this subgroup have the general properties specified for the Organic order and the Fibrisol great group. They differ from Typic Fibrisols by having both a terric layer beneath the surface tier and a subdominant mesic layer thicker than 25 cm in the control section. They may also have cumulic or limnic layers but lack humic and hydric layers.

Terric Humic Fibrisol

Common horizon sequence: Of or Om, <u>Of</u>, <u>Oh</u>, <u>C</u>

Soils of this subgroup have the general properties specified for the Organic order and the Fibrisol great group. They differ from Typic Fibrisols by having both a terric layer beneath the surface tier and a subdominant humic layer thicker than 12 cm in the control section. They may also have mesic, cumulic, or limnic layers but lack a hydric layer.

Hydric Fibrisol

Common horizon sequence: Of or Om, <u>Of</u>, <u>W</u>

Soils of this subgroup have the general properties specified for the Organic order and the Fibrisol great group. They differ from Typic Fibrisols by having a hydric layer (a layer of water that extends from a depth of not less than 40 cm to a depth of more than 1.6 m). They may also have mesic, humic, cumulic, terric, or limnic layers.

Mesisol

Soils of this great group are at a stage of decomposition intermediate between Fibrisols and Humisols. Mesisols have a dominantly mesic middle tier or middle and surface tiers if a terric, lithic, or hydric contact occurs in the middle tier. A mesic layer is an organic layer that fails to meet the requirements of either a fibric or a humic layer. Thus it has a rubbed fiber content ranging from 10% to less than 40% by volume and has a pyrophosphate index of >3 and <5. Mesic material is usually classified on the von Post scale of decomposition as class 5 or 6.

Typic Mesisol

Common horizon sequence: Of, Om or Oh, <u>Om</u>

Soils of this subgroup have the general properties specified for the Organic order and the Mesisol great group. They are composed mainly of organic materials at an intermediate stage of decomposition.

They are identified by the following properties:

1. If present, the middle and bottom tiers are dominantly mesic. A lithic contact may occur.
2. These soils do not have terric, hydric, cumulic, or limnic layers within the middle tier.
3. They lack subdominant humic or fibric layers with a total thickness greater than 25 cm in the middle and bottom tiers or in the middle and surface tiers if a lithic contact occurs in the middle tier.

Fibric Mesisol

Common horizon sequence: Of, Om or Oh, <u>Om</u>, <u>Of</u>, Om

Soils of this subgroup have the general properties specified for the Organic order and the Mesisol great group. They differ from Typic Mesisols by having a subdominant fibric layer thicker than 25 cm in the middle or bottom tiers. These soils do not have a subdominant humic layer thicker than 25 cm.

Humic Mesisol

Common horizon sequence: Of, Om or Oh, <u>Om</u>, <u>Oh</u>, Om

Soils of this subgroup have the general properties specified for the Organic order and the Mesisol great group. They differ from Typic Mesisols by having a subdominant humic layer thicker than 25 cm in the middle or bottom tiers. They may also have a subdominant fibric layer.

Limnic Mesisol

Common horizon sequence: Of, Om or Oh, <u>Om</u>, <u>Oco</u>, Om

Soils of this subgroup have the general properties specified for the Organic order and the Mesisol great group. They differ from Typic Mesisols by having a limnic layer beneath the surface tier. A limnic layer is a layer or layers at least 5 cm thick of coprogenous earth (sedimentary peat), diatomaceous earth, or marl. Also they may have fibric, humic, and cumulic layers but do not have terric or hydric layers.

Cumulic Mesisol

Common horizon sequence: Of, Om or Oh, Om, C, Om

Soils of this subgroup have the general properties specified for the Organic order and the Mesisol great group. They differ from Typic Mesisols by having a cumulic layer beneath the surface tier. They may also have fibric or humic layers but lack terric, hydric, and limnic layers. A cumulic layer consists either of multiple layers of mineral material (alluvium) that together are more than 5 cm thick, or of one layer 5–30 cm thick.

Terric Mesisol

Common horizon sequence: Of, Om or Oh, Om, C, Om

Soils of this subgroup have the general properties specified for the Organic order and the Mesisol great group. They differ from Typic Mesisols by having a terric layer (an unconsolidated mineral layer at least 30 cm thick) beneath the surface tier. They may also have cumulic or limnic layers, but they do not have fibric, humic, or hydric layers within the control section.

Terric Fibric Mesisol

Common horizon sequence: Of, Om or Oh, Om, Of, C, Om

Soils of this subgroup have the general properties specified for the Organic order and the Mesisol great group. They differ from Typic Mesisols by having both a terric layer (an unconsolidated mineral layer at least 30 cm thick) beneath the surface tier and a subdominant fibric layer thicker than 25 cm in the control section. These soils may also have cumulic and limic layers but lack humic and hydric layers.

Terric Humic Mesisol

Common horizon sequence: Of, Om or Oh, Om, Oh, C, Om

Soils of this subgroup have the general properties specified for the Organic order and the Mesisol great group. They differ from Typic Mesisols by having both a terric layer (an unconsolidated mineral layer at least 30 cm thick) beneath the surface tier and a subdominant humic layer thicker than 25 cm within the control section. These soils may also have fibric, cumulic, or limnic layers but lack a hydric layer.

Hydric Mesisol

Common horizon sequence: Of, Om or Oh, Om, W

Soils of this subgroup have the general properties specified for the Organic order and the Mesisol great group. They differ from Typic Mesisols by having a hydric layer (a layer of water that extends from a depth of not less than 40 cm to a depth of more than 1.6 m). These soils may also have fibric, humic, cumulic, terric, or limnic layers.

Humisol

Soils of this great group are at the most advanced stage of decomposition of the great groups of Organic soils. Most of the material is humified with few recognizable fibers. Humisols have a dominantly humic middle tier or middle and surface tiers if a terric, lithic, or hydric contact occurs in the middle tier. A humic layer is an organic layer having less than 10% rubbed fiber by volume and a pyrophosphate index of 3 or less. It has a higher bulk density, usually >0.195 Mg m^{-3}, and a lower water-holding capacity than fibric or mesic layers. Humic material usually is classified on the von Post scale of decomposition as class 7 or higher and rarely in class 6.

Only minor areas of Humisols are known to occur in Canada.

Typic Humisol

Common horizon sequence: Om or Oh, Oh

Soils of this subgroup have the general properties specified for the Organic order and the Humisol great group. They are composed dominantly of well-decomposed organic materials.

They are identified by the following properties:

1. The middle and bottom tiers, if present, are dominantly humic. A lithic contact may occur.
2. They do not have terric, hydric, cumulic, or limnic layers within the middle tier.
3. They have neither subdominant fibric layers with a total thickness greater than 12 cm nor subdominant mesic layers with a total thickness greater than 25 cm in the middle or bottom tiers (or in the middle and surface tiers if a lithic contact occurs in the middle tier).

Fibric Humisol

Common horizon sequence: Om or Oh, <u>Oh</u>, <u>Of</u>, Oh

Soils of this subgroup have the general properties specified for the Organic order and Humisol great group. They differ from Typic Humisols by having a subdominant fibric layer thicker than 12 cm in the middle or bottom tiers. They may also have a subdominant mesic layer.

Mesic Humisol

Common horizon sequence: Om or Oh, <u>Oh</u>, <u>Om</u>, Oh

Soils of this subgroup have the general properties specified for the Organic order and the Humisol great group. They differ from Typic Humisols by having a subdominant mesic layer thicker than 25 cm in the middle or bottom tiers. They lack a subdominant fibric layer below the surface tier.

Limnic Humisol

Common horizon sequence: Om or Oh, <u>Oh</u>, <u>Oco</u>, Oh

Soils of this subgroup have the general properties specified for the Organic order and the Humisol great group. They differ from Typic Humisols by having a limnic layer beneath the surface tier. A limnic layer is a layer or layers at least 5 cm thick of coprogenous earth (sedimentary peat), diatomaceous earth, or marl. They may also have fibric, mesic and cumulic layers but do not have terric or hydric layers.

Cumulic Humisol

Common horizon sequence: Om or Oh, <u>Oh</u>, <u>C</u>, Oh

Soils of this subgroup have the general properties specified for the Organic order and the Humisol great group. They differ from Typic Humisols by having a cumulic layer beneath the surface tier. Also they may have fibric or mesic layers but lack terric, hydric, and limnic layers. A cumulic layer consists either of multiple layers of mineral material (alluvium) that together are more than 5 cm thick, or of one layer 5–30 cm thick.

Terric Humisol

Common horizon sequence: Om or Oh, <u>Oh</u>, <u>C</u>, Oh

Soils of this subgroup have the general properties specified for the Organic order and the Humisol great group. They differ from Typic Humisols by having a terric layer (an unconsolidated mineral layer at least 30 cm thick) beneath the surface tier. They may also have cumulic or limnic layers but do not have fibric, mesic or hydric layers within the control section.

Terric Fibric Humisol

Common horizon sequence: Of or Oh, <u>Oh</u>, <u>Of</u>, <u>C</u>, Oh

Soils of this subgroup have the general properties specified for the Organic order and the Humisol great group. They differ from Typic Humisols by having both a terric layer (an unconsolidated mineral layer at least 30 cm thick) beneath the surface tier and a subdominant fibric layer thicker than 12 cm within the control section. They may also have mesic, cumulic or limnic layers but lack a hydric layer.

Terric Mesic Humisol

Common horizon sequence: Om or Oh, <u>Oh</u>, <u>Om</u>, C, Oh

Soils of this subgroup have the general properties specified for the Organic order and the Humisol great group. They differ from Typic Humisols by having both a terric layer (an unconsolidated mineral layer at least 30 cm thick) beneath the surface tier and a subdominant mesic layer thicker than 25 cm within the control section. These soils may also have cumulic or limnic layers but lack a subdominant fibric or hydric layer.

Hydric Humisol

Common horizon sequence: Om or Oh, <u>Oh</u>, <u>W</u>

Soils of this subgroup have the general properties specified for the Organic order and the Humisol great group. They differ from Typic Humisols by having a hydric layer. These soils may also have fibric, mesic, humic, cumulic, terric, or limnic layers. A hydric layer is a layer of water that extends from a depth of not less than 40 cm to a depth of more than 160 cm.

Folisol

Soils of the Folisol great group are composed of upland organic (folic) materials, generally of forest origin, that are either 40 cm or more in thickness, or are at least 10 cm thick if overlying bedrock or fragmental material. Deep Folisols (greater than 40 cm of folic material) occur frequently in cool, moist, and humid forest ecosystems, particularly on the West Coast of Canada. They also develop in northern regions where soil temperatures are low, but the soil is without permafrost. Shallow Folisols are found throughout Canada and commonly occur on upper slope shedding positions over bedrock and on, or incorporated in, fragmental or skeletal material.

Folic materials are formed under ecosystems different from those of peat materials. Folic materials are the product of upland ecosystem development, whereas peat materials are the product of wetland development.

Folisols are well to imperfectly drained, although they may become saturated after rainfall or snowmelt. They contain organic C at a level of >17% (about 30% or more organic matter) by weight in diagnostic horizons. Folic materials qualify as Folisols if they meet the following criteria:

1. Folic material is 40 cm or more in depth; or

2. Folic material is 10 cm or more in depth if directly overlying a lithic contact or fragmental material, or if occupying voids in fragmental or skeletal material; or

3. Folic material is more than twice the thickness of a mineral soil layer if the mineral layer is less than 20 cm thick.

---

[1] Various mineral soil horizons.

Folic materials containing permafrost at depths of 1 m or less are classified as Cryosolic soils.

The Folisol great group is divided into four subgroups, based on the degree of decomposition of the folic material (as distinguished by the diagnostic F and H soil horizons) or on the type of organic materials in the control section, or on both. Layers or pockets of decaying wood may be designated as an F or an H horizon, and some soil pedons have significant amounts of this type of folic material (*see* Lignic Folisol). The L horizon, if present, is considered to be parent material.

Hemic Folisol

Common horizon sequence: L, <u>F</u>, H, O, R, (M[1])

Soils of this subgroup are composed dominantly of the moderately decomposed F horizon in the control section and may have subdominant H and O horizons, each less than 10 cm thick. They commonly have a lithic contact or fragmental layers but meet the requirements of the Folisol great group. The F horizon consists of partly decomposed folic material generally derived from mosses, leaves, twigs, reproductive structures, and woody materials containing numerous live and dead roots.

Hemic Folisols usually occur on upper slope shedding positions and commonly consist of shallow folic material over bedrock or fragmental material, or the folic materials may occupy voids in fragmental material. There may be a thin layer of mineral soil separating the folic horizon from bedrock or from the fragmental material.

Humic Folisol

Common horizon sequence: L, F, <u>H</u>, O, R, (M[1])

Soils of this subgroup are composed dominantly of the well-decomposed H horizon in the control section and may have subdominant F and O horizons each less than 10 cm thick. A lithic contact, fragmental, or mineral layers may be common in the control section, but the soils meet the requirements for the Folisol great group.

Humic Folisols occur most frequently in cool, moist, humid forest ecosystems. Although they occur in many landscape positions, they commonly develop on lower slopes and in valley bottoms. Rooting channels and other voids are common in these soils.

Lignic Folisol

Common horizon sequence: L, F, H, R, (M[1])

Soils of this subgroup are dominated by F or H horizons, which are composed primarily of moderately to well-decomposed woody materials. These materials occupy more than 30% of the surface area of the F and H horizons. The decaying wood that makes up the F and H horizons generally consists of trees that have been blown down. The destruction of trees is either a continuing process in unevenly aged forests or occurs at periodic intervals as a result of major storms, which create evenly aged forest stands. Lignic Folisols also occur in a second-growth forest as a result of logging.

Histic Folisol

Common horizon sequence: L, F, H, O, R, (M[1])

Soils of this subgroup are dominated by F or H horizons and are directly underlain by a significant (greater than 10 cm) O horizon. Generally, saturation or high water tables (resulting from drainage impediment caused by mineral horizon cementation or localized bedrock configuration) initially encouraged the production of peat. Peat development subsequently became deep enough to produce surface conditions suitable for forest encroachment and Folisol development.

---

[1] Various mineral soizl horizons.

Chapter 10

Podzolic Order

Great Group	Subgroup
Humic Podzol	Orthic Humic Podzol O.HP
	Ortstein Humic Podzol OT.HP
	Placic Humic Podzol P.HP
	Duric Humic Podzol DU.HP
	Fragic Humic Podzol FR.HP
Ferro-Humic Podzol	Orthic Ferro-Humic Podzol O.FHP
	Ortstein Ferro-Humic Podzol OT.FHP
	Placic Ferro-Humic Podzol P.FHP
	Duric Ferro-Humic Podzol DU.FHP
	Fragic Ferro-Humic Podzol FR.FHP
	Luvisolic Ferro-Humic Podzol LU.FHP
	Sombric Ferro-Humic Podzol SM.FHP
	Gleyed Ferro-Humic Podzol GL.FHP
	Gleyed Ortstein Ferro-Humic Podzol GLOT.FHP
	Gleyed Sombric Ferro-Humic Podzol GLSM.FHP
Humo-Ferric Podzol	Orthic Humo-Ferric Podzol O.HFP
	Ortstein Humo-Ferric Podzol OT.HFP
	Placic Humo-Ferric Podzol P.HFP
	Duric Humo-Ferric Podzol DU.HFP
	Fragic Humo-Ferric Podzol FR.HFP
	Luvisolic Humo-Ferric Podzol LU.HFP
	Sombric Humo-Ferric Podzol SM.HFP
	Gleyed Humo-Ferric Podzol GL.HFP
	Gleyed Ortstein Humo-Ferric Podzol GLOT.HFP
	Gleyed Sombric Humo-Ferric Podzol GLSM.HFP

A diagrammatic representation of profiles of some subgroups of the Podzolic order is shown in Figure 37. Individual subgroups may include soils that have horizon sequences different from those shown. In the description of each subgroup, presented later in this chapter, a common horizon sequence is given; diagnostic horizons are underlined and some other commonly occurring horizons are listed.

Soils of the Podzolic order have B horizons in which the dominant accumulation product is amorphous material composed mainly of humified organic matter combined in varying degrees with Al and Fe. Typically Podzolic soils occur in coarse- to medium-textured, acid parent materials, under forest or heath vegetation in cool to very cold humid to perhumid climates. However, some occur under soil environmental conditions outside this range. For example, minor areas of Podzolic soils occur in wet sandy sites in areas of subhumid climate. Other Podzolic soils have formed in parent materials that were once calcareous.

Podzolic soils can usually be recognized readily in the field. Generally they have organic surface horizons that are commonly L, F, and H but may be Of or Om and have a light-colored eluvial horizon, Ae, which may be absent. Most Podzolic soils have a reddish brown to black B horizon with an abrupt upper boundary and lower B or BC horizons with colors that become progressively yellower in hue and lower in chroma with depth, except in reddish-colored parent materials.

Soils of the Podzolic order are defined based on a combination of morphological and chemical criteria of the B horizons. Soils of the order must meet all the following morphological limits and those specified under either 1 or 2 of the chemical limits.

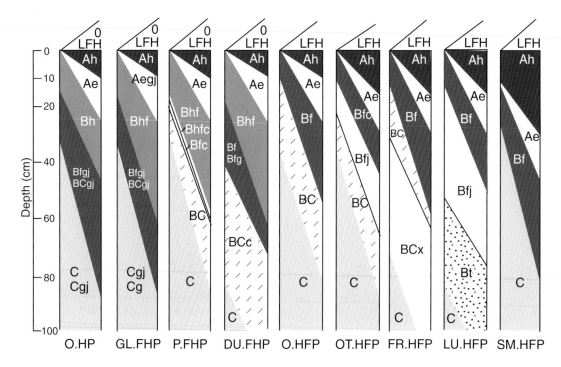

Figure 37 Diagrammatic horizon pattern of some subgroups of the Podzolic order.

Morphological Limits

1. The podzolic B horizon is at least 10 cm thick and has moist, crushed colors as follows:

   a. The color is black or the hue is either 7.5YR or redder or 10YR near the upper boundary and becomes yellower with depth.

   b. The chroma is higher than 3 or the value is 3 or less.

2. The accumulation of amorphous material in the podzolic B horizon is indicated by:

   a. Brown to black coatings on some mineral grains or brown to black microaggregates.

   b. A silty feel when rubbed wet unless the material is cemented.

3. The texture of the podzolic B horizon is coarser than clay.

4. The soil either has no Bt horizon or the upper boundary of the Bt horizon is at a depth greater than 50 cm from the mineral soil surface.

Chemical Limits

1. The soils have a B subhorizon (Bh) at least 10 cm thick, with both color value and chroma (moist) of 3 or less, that contains more than 1% organic C, less than 0.3% pyrophosphate-extractable Fe, and has a ratio of organic C to pyrophosphate-extractable Fe of 20 or more.

2. The soils have a B subhorizon (Bf or Bhf) at least 10 cm thick with the following characteristics:

   a. An organic C content of 0.5% or more.

   b. A pyrophosphate-extractable Fe+Al content of 0.6% or more in textures finer than sand and of 0.4% or more in sands (coarse sand, sand, fine sand, and very fine sand).

   c. A ratio of pyrophosphate-extractable Fe+Al to clay (≤0.002 mm) of more than 0.05.

   d. A ratio of organic C to pyrophosphate-extractable Fe of less than 20, or pyrophosphate-extractable Fe of at least 0.3% or either color value or chroma of more than 3.

A Bf horizon contains 0.5–5% organic C, and a Bhf horizon contains more than 5% organic C.

Any B horizon that satisfies the specified morphological and chemical requirements is a podzolic B horizon. In the following cases, the color criteria of a podzolic B horizon are waived for the following: Ap horizons that meet the chemical limits; and B subhorizons that meet the chemical limits specified. Some

Bh, Bhf, and Bf horizons do not qualify as podzolic B horizons because they are too thin. To determine whether a pedon meets Podzolic order criteria, it is necessary to sample only the 10 cm of B horizon that are most strongly expressed. If all of the B horizon looks the same, it may be desirable to take three samples (top, middle, and bottom 10 cm). Similarly, to determine whether a pedon has a Bhf or a Bh horizon, it is necessary to sample a subhorizon only 10 cm thick that appears most likely to meet the limits of Bhf or Bh. Average samples should consist of 10 cm of subhorizon taken uniformly from an exposure 50 cm wide or wider. For many studies it is desirable to sample all subhorizons.

Some soils that are not Podzolic will satisfy the minimum morphological limits specified. However, these limits are thought to be useful to exclude from the order certain soils having horizons that satisfy the chemical limits specified but otherwise do not resemble Podzolic soils. To be classified as Podzolic, a borderline soil must meet both the morphological and the chemical limits specified.

Some acid Ah horizons satisfy the morphological and chemical criteria of podzolic B horizons. These are commonly associated with volcanic ash. No specific criteria have been developed to distinguish these horizons from Bhf or Bf horizons. The following guidelines are useful:

1. Such Ah horizons are generally black and underlain by brown or dark brown B horizons. For example, the moist color of the A horizon may be 10YR 2/2 and that of the B 7.5YR 4/4.

2. The ratio of humic to fulvic acid in these Ah horizons is greater than 1:2 (usually 1:1 or higher) and in the underlying B horizon it is less than 1:2.

3. Less than 50% of the total organic C is extracted from these A horizons by alkali pyrophosphate and more than 50% is extracted from the associated B horizons.

Some associated properties of podzolic B horizons

Besides the properties specified as diagnostic, podzolic B horizons have a number of associated properties that may be useful in distinguishing them from other B horizons. Some of these associated properties are as follows:

1. They have a high pH-dependent cation exchange capacity (CEC). The difference (ΔCEC) between CEC measured at pH 7 and at the pH of the soil is usually at least 8 cmol kg^{-1}.

2. They have a base saturation, as determined by an unbuffered salt, nearly always below 80% and commonly less than 50%.

3. Unless cemented, they have a higher water-holding capacity than non Bf horizons of similar texture.

4. They have a high capacity to fix phosphate.

5. Although they commonly contain appreciably more clay than the overlying Ae horizon (if present), usually very little of the clay occurs as oriented coatings on particles or peds.

6. They give a strongly alkaline reaction in NaF. As a field test a 2% suspension of soil in 1 M NaF gives a pH of 10.3 or more for most podzolic B horizons. Volcanic ash samples also commonly give a high pH with this test, thus the NaF test is not useful for identifying podzolic B horizons in materials containing volcanic ash.

Distinguishing Podzolic Soils from Soils of Other Orders

Guidelines for distinguishing Podzolic soils from soils of other orders with which they might be confused are as follows:

Luvisolic Some Podzolic and some Luvisolic soils have Ae, Bf, and Bt horizons. These soils are classified as Podzolic if the upper boundary of the Bt horizon is at a depth below 50 cm and as Luvisolic if it is at a depth of 50 cm or less.

Brunisolic In the continuum of soils in nature many pedons have properties close to the arbitrary boundary line between Podzolic soils and acid Brunisolic soils. If the B horizon meets the requirements of a podzolic B, the soils are classified as Podzolic.

Gleysolic A podzolic B horizon takes precedence over gley features. Thus a soil having both a podzolic B horizon and gley colors as specified for soils of the Gleysolic order is classified as Podzolic.

Organic Some soils have podzolic B horizons underlying a thick layer of peat or folic materials. The soil is classified as Organic if the peat layer is either 60 cm or more in depth and consists of fibric materials, or 40 cm or

	Humic Podzol	Ferro-Humic Podzol	Humo-Ferric Podzol
B horizon	Bh ≥ 10 cm thick	Bhf ≥ 10 cm thick	Bf, or thin Bhf+Bf ≥ 10 cm thick
Organic C	>1%	>5%	= 0.5–5%
Other	pyrophosphate Fe <0.3%, pyrophosphate Fe ≥20	pyrophosphate Al+Fe ≥0.6% (≥0.4% for sands)	pyrophosphate Al+Fe (≥0.4% for sands)

more in depth if it consists of mesic, humic, or folic materials.

The Podzolic order is divided into three great groups: Humic Podzol, Ferro-Humic Podzol, and Humo-Ferric Podzol based on the organic C content and the organic C to pyrophosphate-extractable Fe ratio of the podzolic B horizon as shown in the Pozolic order chart.

Subgroups are separated based on the kind and sequence of the horizons indicating conformity with the central concept of the great group, the presence of additional horizons, or intergrading to soils of other orders. Some features formerly recognized taxonomically at the subgroup level are now recognized at either the family (lithic, some cryic) or series (turbic) level. These features may also be recognized as phases of subgroups, great groups, or orders.

Humic Podzol

These soils have a dark-colored podzolic B horizon that contains very little extractable Fe. They occur typically in wet sites so that they are saturated with water during some periods of the year. Characteristically they occur under heath, forest and heath, sphagnum, or western coastal forest vegetation in maritime fringe environments, on some sites at high elevations inland, and in peaty depressions. Under virgin conditions Humic Podzols usually have thick L, F, and H or O horizons underlain by a light-colored eluvial horizon (Ae), an eluvial horizon darkened by humic material, or by a podzolic B horizon, which is usually a Bh. The B horizon may include several kinds of podzolic B subhorizons: Bh, Bhf, and Bf, which may be cemented (ortstein, placic) or friable. The material below the podzolic B horizon may be cemented (duric), compact and brittle (fragipan), or friable.

Humic Podzols have a Bh horizon at least 10 cm thick that usually occurs at the top of the B horizon but may occur below other B horizons. The Bh horizon contains more than 1% organic C and less than 0.3% pyrophosphate-extractable Fe and has a ratio of organic C to pyrophosphate-extractable Fe of 20 or more.

Humic Podzols are generally strongly acid and their B horizons are usually less than 50% base saturated (neutral salt). The pH dependent CEC of the Bh horizon is usually well above 8 cmol kg^{-1}.

Under disturbed conditions and where the Bh horizon directly underlies the organic surface layer, the Bh may be confused with an Ah horizon. The guidelines that aid in making this distinction are that more than 50% of the organic C of Bh horizons is extractable by NaOH-Na$_4$P$_2$O$_7$ and more than 50% of the extractable C of Bh horizons is fulvic acid carbon. Cultivated Humic Podzols are identified by properties of the B horizon below the cultivated layer.

Distinguishing Bh from Bhf horizons may be a problem in the field. The following two guidelines are useful:

- Bh horizon material usually does not turn redder on ignition because of its low Fe content.
- Bh horizons are generally almost black. However, some Bhf horizons also have chromas of only 1 or 2.

Humic Podzols are divided into five subgroups based on the kind and sequence of the horizons. No gleyed subgroup has been included because Humic Podzols usually occur in wet sites; thus the great group implies some degree of gleying.

Orthic Humic Podzol

Common horizon sequence: O or LFH, Ae, Bh, Bfgj, BCgj, Cg

These soils have the general properties specified for the Podzolic order and the

Humic Podzol great group. They are identified by the following properties:

1. They have a Bh horizon at least 10 cm thick.

2. They do not have an ortstein horizon at least 3 cm thick, a placic horizon, a duric horizon, or a fragipan.

Usually Orthic Humic Podzols have L, F, and H or O horizons and an Ae horizon. Commonly they have a Bhf or Bf horizon underlying the Bh horizons. They may have mottling that indicates gleying at any depth within the control section. Parts of the Bhf or Bf may be cemented but do not meet the requirements of an ortstein horizon.

Ortstein Humic Podzol

Common horizon sequence: LFH or O, Ae, <u>Bh</u> or <u>Bhc</u>, <u>Bfc</u>, Cgj

These soils have the general properties specified for the Podzolic order and the Humic Podzol great group. They differ from Orthic Humic Podzols by having an ortstein horizon at least 3 cm thick. An ortstein horizon is a Bh, Bhf, or Bf horizon that is strongly cemented and occurs in at least one-third of the lateral extent of the pedon. The ortstein horizon is designated as Bhc, Bhfc, or Bfc depending upon its organic C and extractable Fe content. Ortstein horizons are generally reddish brown to very dark reddish brown in color. Usually Ortstein Humic Podzols have L, F, and H or O horizons and an Ae horizon. They may have mottling that indicates gleying at any depth within the control section and placic or duric horizons or a fragipan.

Placic Humic Podzol

Common horizon sequence: LFH or O, Ae, <u>Bh</u>, <u>Bhfc</u> or <u>Bfc</u>, BCgj, Cgj

These soils have the general properties specified for the Podzolic order and the Humic Podzol great group. They differ from Orthic Humic Podzols by having a placic horizon within the control section. A placic horizon (Bhfc, Bfc, Bfgc) consists of a single thin layer that is commonly 5 mm or less in thickness, or a series of thin layers that are irregular or involute, hard, impervious, often vitreous, and dark reddish brown to black in color. These thin horizons are apparently cemented by Fe-organic complexes, hydrated Fe oxides, or a mixture of Fe and Mn oxides. The placic horizon or thin iron pan may occur

in any part of the B horizon except the Bh, and commonly it extends into the BC horizon.

Placic Humic Podzols usually have L, F, and H or O horizons and an Ae horizon. They do not have an ortstein horizon but may have a duric horizon or a fragipan. Evidence of gleying in the form of dull colors or mottling is commonly apparent especially above depressions in the placic horizon. These soils occur most commonly in wet sites in maritime regions; frequently the surface is peaty.

Duric Humic Podzol

Common horizon sequence: LFH or O, Ae, <u>Bh</u>, Bhf, <u>BCc</u>, Cgj

These soils have the general properties specified for the Podzolic order and the Humic Podzol great group. They differ from Orthic Humic Podzols by having a duric horizon within the control section. A duric horizon is a strongly cemented horizon that does not satisfy the criteria of a podzolic B horizon. It usually has an abrupt upper boundary to an overlying podzolic B horizon and a diffuse lower boundary at least 50 cm below. Cementation is usually strongest near the upper boundary, which occurs commonly at a depth of 40–80 cm from the mineral surface. Usually the color of a duric horizon differs little from that of the parent material, and the structure is usually massive or very coarse platy. Moist clods at least 3 cm thick usually cannot be broken in the hands. Air-dry clods of the material do not slake when immersed in water. Some duric horizons may meet the requirements of a Bt horizon (Btc).

Duric Humic Podzols usually have L, F and H or O horizons. They have neither an ortstein nor a placic horizon but may have mottles that indicate gleying in some part of the control section.

Fragic Humic Podzol

Common horizon sequence: LFH or O, Ae, <u>Bh</u>, Bf, <u>BCxgj</u>, Cgj

These soils have the general properties specified for the Podzolic order and the Humic Podzol great group. They differ from Orthic Humic Podzols by having a fragipan within the control section. A fragipan (Bx or BCx) is a subsurface horizon of high bulk density that has firm and brittle consistence when moist and hard to extremely hard consistence when dry. Usually it is of medium texture. Commonly it has bleached fracture planes separating very coarse prismatic units and the secondary structure is platy. Usually the fragipan is similar in color to the parent material, but it differs in structure and consistence and sometimes in bulk density. The upper boundary of a fragipan is usually either abrupt or clear, but the lower boundary is usually diffuse. Commonly it is necessary to dig to about 3 m to expose clearly the material beneath the lower boundary of the fragipan. Air-dry clods of fragipans slake in water. A fragipan may have clay skins and meet the limits of a Bt horizon (Btx).

Fragic Humic Podzols usually have L, F, and H or O horizons and an Ae horizon. They do not have ortstein, placic, or duric horizons but may have mottles that indicate gleying at some depth within the control section.

Ferro-Humic Podzol

These soils have a dark-colored podzolic B horizon with a high content of organic C and an appreciable amount of extractable Fe and Al. They occur typically in the more humid part of the region of Podzolic soils under forest vegetation, or forest with heath or moss undercover. Under virgin conditions these soils usually have thick L, F, and H or O horizons; they may have an Ah horizon and usually have a light-colored Ae horizon. The podzolic B horizon is usually thick and dark reddish brown in the upper part and grades to lighter colors of higher chroma with depth. Typically the Bhf horizon is of higher chroma (commonly 2, 3, or 4) than the Bh horizon of Humic Podzols (commonly 1 or 2). The material below the podzolic B horizon may be cemented (duric), compact and brittle (fragipan), or friable.

Ferro-Humic Podzols have a Bhf horizon at least 10 cm thick and lack a Bh horizon at least 10 cm thick. The Bhf horizon contains 5% or more organic C and 0.6% or more pyrophosphate-extractable Fe+Al (0.4% for sands). It has a ratio of organic C to pyrophosphate-extractable Fe of less than 20, or has 0.3% or more pyrophosphate-extractable Fe, or both.

Ferro-Humic Podzols are generally both strongly acid and less than 50% base saturated (neutral salt). The pH-dependent CEC of the Bhf horizon is usually well above 8 cmol kg^{-1} and is commonly 25 cmol kg^{-1} or more. The Bhf horizon of these soils usually has a markedly silty feel when rubbed moist, which is presumably because of its high content of amorphous material.

Ferro-Humic Podzols are divided into ten subgroups based on the kind and sequence of horizons. Gleyed subgroups are not differentiated for those soils having a relatively impermeable subsoil horizon (placic, duric, fragic, Bt). A Gleyed Ortstein subgroup is included because some ortstein horizons are permeable to water and do not result in temporary gleying.

Orthic Ferro-Humic Podzol

Common horizon sequence: LFH or O, Ae, Bhf Bf, BC, C

These soils have the general properties specified for the Podzolic order and the Ferro-Humic Podzol great group. They are identified by the following properties:

1. They have a Bhf horizon at least 10 cm thick.
2. They do not have a Bh horizon at least 10 cm thick, an ortstein horizon at least 3 cm thick, a placic horizon, a duric horizon, a fragipan, a Bt horizon, an Ah horizon at least 10 cm thick, or evidence of gleying in the form of distinct or prominent mottling within 1 m of the surface.

Usually Orthic Ferro-Humic Podzols have L, F, and H or O horizons and an Ae horizon. Commonly they have a Bf horizon underlying the Bhf. Parts of the Bhf or Bf may be cemented but do not meet the requirements of an ortstein horizon.

Ortstein Ferro-Humic Podzol

Common horizon sequence: LFH or O, Ae, Bhf, Bhfc or Bfc, BC, C

These soils have the general properties specified for the Podzolic order and the Ferro-Humic Podzol great group. They differ from Orthic Ferro-Humic Podzols by having an ortstein horizon at least 3 cm thick. An ortstein horizon is a Bh, Bhf, or Bf horizon that is strongly cemented and occurs in at least one-third of the lateral extent of the pedon. The ortstein horizon in this subgroup is designated as Bhfc or Bfc depending upon its organic C content and is generally reddish brown to very dark reddish brown in color. Usually Ortstein Ferro-Humic Podzols have L, F, and H or O horizons and an Ae horizon. They may have one or more of placic, duric, Ah, or Bt horizons, a fragipan, or faint mottling.

Placic Ferro-Humic Podzol

Common horizon sequence: LFH or O, Ae, <u>Bhf</u>, <u>Bhfc</u> or <u>Bfc</u>, Bf, BC, C

These soils have the general properties specified for the Podzolic order and the Ferro-Humic Podzol great group. They differ from Orthic Ferro-Humic Podzols by having a placic horizon within the control section. A placic horizon (Bhfc, Bfc, Bfgc) consists of a single thin layer (commonly 5 mm or less thick) or a series of thin layers that are irregular or involute, hard, impervious, often vitreous, and dark reddish brown to black. These thin horizons are apparently cemented by Fe-organic complexes, hydrated Fe oxides, or a mixture of Fe and Mn oxides. The placic horizon, or thin iron pan, may occur in any part of the B horizon and commonly extends into the BC horizon.

Placic Ferro-Humic Podzols usually have L, F, and H or O horizons and an Ae horizon. They do not have an ortstein horizon but may have a duric, Ah, or Bt horizon, or a fragipan. Evidence of gleying in the form of dull colors or mottling is commonly apparent especially above depressions in the placic horizon. These soils occur most commonly in coarse textured deposits in perhumid maritime climates.

Duric Ferro-Humic Podzol

Common horizon sequence: LFH or O, Ae, <u>Bhf</u>, <u>BCc</u>, C

These soils have the general properties specified for the Podzolic order and the Ferro-Humic Podzol great group. They differ from Orthic Ferro-Humic Podzols by having a duric horizon within the control section. A duric horizon is a strongly cemented horizon that does not satisfy the criteria of a podzolic B horizon. It usually has an abrupt upper boundary to an overlying podzolic B horizon and a diffuse lower boundary at least 50 cm below. Cementation is usually strongest near the upper boundary, which occurs commonly at a depth of 40–80 cm from the mineral surface. Usually the color of a duric horizon differs little from that of the parent material. As well the structure is usually massive or very coarse platy. Moist clods at least 3 cm thick usually cannot be broken in the hands. Air-dry clods of the material do not slake when immersed in water. Some duric horizons may meet the requirements of a Bt horizon (Btc).

Duric Ferro-Humic Podzols usually have L, F, and H or O horizons. They do not have an ortstein or a placic horizon but may have an Ah horizon and mottles that indicate gleying in some part of the control section. These soils occur most commonly in coastal southwestern British Columbia.

Fragic Ferro-Humic Podzol

Common horizon sequence: LFH or O, Ae, <u>Bhf</u>, Bf, <u>BCx</u>, C

These soils have the general properties specified for the Podzolic order and the Ferro-Humic Podzol great group. They differ from Orthic Ferro-Humic Podzols by having a fragipan within the control section. A fragipan (Bx or BCx) is a subsurface horizon of high bulk density that has firm and brittle consistence when moist and hard to extremely hard consistence when dry. Usually it is of medium texture. Commonly it has bleached fracture planes separating very coarse prismatic units and the secondary structure is platy. The fragipan is usually similar in color to the parent material, but it differs in structure and consistence and sometimes in bulk density. The upper boundary of a fragipan is usually either abrupt or clear, but the lower boundary is usually diffuse. It is often necessary to dig to about 3 m to expose clearly the material beneath the lower boundary of the fragipan. Air-dry clods of fragipans slake in water. A fragipan may have clay skins and meet the limits of a Bt horizon (Btx).

Fragic Ferro-Humic Podzols usually have L, F, and H or O horizons and an Ae horizon. They do not have ortstein, placic, or duric horizons but may have an Ah horizon and mottles that indicate gleying at some depth within the control section.

Luvisolic Ferro-Humic Podzol

Common horizon sequence: LFH or O, Ae, <u>Bhf</u>, Bf, <u>Bt</u>, C

These soils have the general properties specified for the Podzolic order and the Ferro-Humic Podzol great group. They differ from Orthic Ferro-Humic Podzols by having a Bt horizon of which the upper boundary is at a depth of more than 50 cm from the mineral surface. If the upper boundary of the Bt horizon is ≤50 cm from the surface the soil is classified in the Luvisolic order.

Luvisolic Ferro-Humic Podzols usually have L, F, and H or O horizons and an Ae horizon and they may have an Ah horizon. They do not have ortstein, placic, or duric horizons, or a fragipan but may have mottles that indicate gleying at some depth within the control section.

Sombric Ferro-Humic Podzol

Common horizon sequence: LFH or O, <u>Ah</u>, Ae, <u>Bhf</u>, Bf, BC, C

These soils have the general properties specified for the Podzolic order and the Ferro-Humic Podzol great group. They differ from Orthic Ferro-Humic Podzols by having an Ah horizon at least 10 cm thick.

Usually Sombric Ferro-Humic Podzols have L, F, and H or O horizons and may have an Ae horizon but do not have ortstein, placic, duric, or Bt horizons, a fragipan, or distinct or prominent mottles that indicate gleying.

Gleyed Ferro-Humic Podzol

Common horizon sequence: LFH or O, Aegj, <u>Bhf</u>, Bfgj, BCg, Cg

These soils have the general properties specified for the Podzolic order and the Ferro-Humic Podzol great group. They differ from Orthic Ferro-Humic Podzols by having distinct or prominent mottles that indicate gleying within 1 m of the surface. Usually they have thick L, F, and H or O horizons. They do not have ortstein, placic, duric, or Bt horizons, a fragipan, or an Ah horizon at least 10 cm thick.

Gleyed Ortstein Ferro-Humic Podzol

Common horizon sequence: LFH or O, Aegj, <u>Bhf</u>, <u>Bhfc</u> or <u>Bfcg</u>, BCg, Cg

These soils have the general properties specified for the Podzolic order and the Ferro-Humic Podzol great group. They differ from Ortstein Ferro-Humic Podzols by having distinct to prominent mottles that indicate gleying within 1 m of the surface. Usually they have thick L, F, and H or O horizons. They may have placic, duric, Ah, or Bt horizons, or a fragipan.

Gleyed Sombric Ferro-Humic Podzol

Common horizon sequence: LFH or O, <u>Ah</u>, Aegj, <u>Bhf</u>, Bfgj, BCgj, C

These soils have the general properties specified for the Podzolic order and the Ferro-Humic Podzol great group. They differ from Sombric Ferro-Humic Podzols by having distinct or prominent mottles that indicate gleying within 1 m of the surface. Usually they have thick L, F, and H or O horizons but do not have ortstein, placic, duric, or Bt horizons, or a fragipan.

Humo-Ferric Podzol

These soils have a brownish-colored podzolic B horizon with less organic matter than the B horizon of Ferro-Humic Podzols. They occur widely both in less humid sites of the region of Podzolic soils and in humid sites. Typically they occur under coniferous, mixed, and deciduous forest vegetation but may occur under shrub and grass vegetation. Under virgin conditions these soils usually have L, F, and H horizons and may have an Ah horizon. Usually they have a light-colored Ae horizon with an abrupt lower boundary to a podzolic B horizon in which the reddest hues or highest chromas and lowest color values usually occur near the top of the horizon and fade with depth. Typically the Bf horizon of Humo-Ferric Podzols has higher color values and chromas than the Bhf horizon of Ferro-Humic Podzols. Parts of the Podzolic B horizon may be cemented and the underlying material may be cemented (duric), compact and brittle (fragipan), or friable.

Humo-Ferric Podzols have a podzolic B horizon at least 10 cm thick but do not have Bh or Bhf horizons at least 10 cm thick. The podzolic B horizon of Humo-Ferric Podzols may include a thin Bhf subhorizon, but usually it is a Bf horizon only. A Bf horizon contains 0.5–5% organic C and 0.6% or more pyrophosphate-extractable Fe+Al (0.4% for sands). Pyrophosphate-extractable Fe is at least 0.3%, or the ratio of organic C to pyrophosphate extractable Fe is less than 20, or both are true. The ratio of pyrophosphate-extractable Fe+Al to clay is more than 0.05.

Humo-Ferric Podzols are generally strongly acid and less than 50% base saturated (neutral salt). The pH-dependent CEC of the Bf horizon is usually at least 8 cmol kg^{-1}.

Typically the podzolic B horizon of a Humo-Ferric Podzol contains less amorphous material than is characteristic of a Ferro-Humic Podzol.

Humo-Ferric Podzols are divided into ten subgroups based on the kind and sequence of the horizons, the same as Ferro-Humic Podzols.

Orthic Humo-Ferric Podzol

Common horizon sequence: LFH, Ae, Bf, BC, C

These soils have the general properties specified for the Podzolic order and the Humo-Ferric Podzol great group. They are identified by the following properties:

1. They have a podzolic B horizon at least 10 cm thick (Bf or thin Bhf and Bf).
2. They do not have a Bh horizon at least 10 cm thick, a Bhf horizon at least 10 cm thick, an ortstein horizon at least 3 cm thick, a placic horizon, a duric horizon, a fragipan, a Bt horizon, an Ah horizon at least 10 cm thick, nor evidence of gleying in the form of distinct or prominent mottles within 1 m of the surface.

Usually Orthic Humo-Ferric Podzols have L, F, and H or O horizons and an Ae horizon. Parts of the Bf may be cemented, but it does not meet the requirements of an ortstein horizon.

Ortstein Humo-Ferric Podzol

Common horizon sequence: LFH, Ae, Bfc, Bfj, C

These soils have the general properties specified for the Podzolic order and the Humo-Ferric Podzol great group. They differ from Orthic Humo-Ferric Podzols by having an ortstein horizon at least 3 cm thick. An ortstein horizon in this subgroup is a Bhf or Bf horizon that is strongly cemented and occurs in at least one-third of the lateral extent of the pedon. Ortstein horizons are generally reddish brown to very dark reddish brown in color. Usually Ortstein Humo-Ferric Podzols have L, F, and H or O horizons and an Ae horizon. They may have faint mottling, and placic, duric, Ah, or Bt horizons, or a fragipan.

Placic Humo-Ferric Podzol

Common horizon sequence: LFH or O, Ae, Bf, Bfc, BC, C

These soils have the general properties specified for the Podzolic order and the Humo-Ferric Podzol great group. They differ from Orthic Humo-Ferric Podzols by having a placic horizon within the control section. A placic horizon (Bhfc, Bfc, Bfgc) consists of a single thin layer (commonly 5 mm or less in thickness) or a series of thin layers that are irregular or involute, hard, impervious, commonly vitreous, and dark reddish brown to black. These thin horizons are apparently cemented by Fe-organic complexes, hydrated Fe oxides, or a mixture of Fe and Mn oxides. The placic horizon or thin iron pan may occur in any part of the B horizon and commonly extends into the BC horizon.

Placic Humo-Ferric Podzols usually have L, F, and H or O horizons and an Ae horizon. They do not have an ortstein horizon but may have a duric horizon or a fragipan. Evidence of gleying in the form of dull colors or mottling is commonly apparent especially above depressions in the placic horizon. These soils occur most commonly in wet sites in maritime regions; frequently the surface is peaty.

Duric Humo-Ferric Podzol

Common horizon sequence: LFH, Ae, Bf, BCc, C

These soils have the general properties specified for the Podzolic order and the Humo-Ferric Podzol great group. They differ from Orthic Humo-Ferric Podzols by having a duric horizon within the control section. A duric horizon is a strongly cemented horizon that does not satisfy the criteria of a podzolic B horizon. Usually it has an abrupt upper boundary to an overlying podzolic B horizon and a diffuse lower boundary at least 50 cm below. Cementation is usually strongest near the upper boundary, which occurs commonly at a depth of 40–80 cm from the mineral surface. The color of a duric horizon usually differs little from that of the parent material and the structure is usually massive or very coarse platy. Moist clods at least 3 cm thick usually cannot be broken in the hands. Air-dry clods of the material do not slake when immersed in water. Some duric horizons may meet the requirements of a Bt horizon (Btc).

Duric Humo-Ferric Podzols usually have L, F, and H horizons. They have neither an ortstein nor a placic horizon but may have an Ah horizon and mottles that indicate gleying in some part of the control section. These soils occur most commonly in coastal southwestern British Columbia.

Fragic Humo-Ferric Podzol

Common horizon sequence: LFH, Ae, <u>Bf</u>, <u>BCx</u>, C

These soils have the general properties specified for the Podzolic order and the Humo-Ferric Podzol great group. They differ from Orthic Humo-Ferric Podzols by having a fragipan within the control section. A fragipan (Bx or BCx) is a subsurface horizon of high bulk density that is firm and brittle when moist and hard to extremely hard when dry. Usually it is of medium texture. Commonly it has bleached fracture planes separating very coarse prismatic units. The secondary structure is platy. Usually the fragipan has a color similar to that of the parent material but differs in structure and consistence and sometimes in bulk density. The upper boundary of a fragipan is usually either abrupt or clear, but the lower boundary is usually diffuse. Commonly it is necessary to dig to about 3 m to expose clearly the material beneath the lower boundary of the fragipan. Air-dry clods of fragipans slake in water. A fragipan may have clay skins and meet the limits of a Bt horizon (Btx).

Fragic Humo-Ferric Podzols usually have L, F, and H horizons and an Ae horizon. They have neither ortstein, placic, nor duric horizons but may have an Ah horizon and mottles that indicate gleying at some depth within the control section.

Luvisolic Humo-Ferric Podzol

Common horizon sequence: LFH, Ae, <u>Bf</u>, <u>Bt</u>, C

These soils have the general properties specified for the Podzolic order and the Humo-Ferric Podzol great group. They differ from Orthic Humo-Ferric Podzols by having a Bt horizon of which the upper boundary is at a depth of more than 50 cm from the mineral surface. If the upper boundary of the Bt horizon is ≤50 cm from the surface, the soil is classified in the Luvisolic order.

Luvisolic Humo-Ferric Podzols usually have L, F, and H horizons and an Ae horizon. They may also have an Ah horizon. They have neither ortstein, duric, nor placic horizons, nor a fragipan but may have mottles that indicate gleying at some depth within the control section.

Sombric Humo-Ferric Podzol

Common horizon sequence: LFH, <u>Ah</u>, Ae, <u>Bf</u>, BC, C

These soils have the general properties specified for the Podzolic order and the Humo-Ferric Podzol great group. They differ from Orthic Humo-Ferric Podzols by having an Ah horizon at least 10 cm thick.

Usually Sombric Humo-Ferric Podzols have L, F, and H horizons and may have an Ae horizon. They have neither ortstein, placic, duric, nor Bt horizons, nor a fragipan, nor distinct or prominent mottles that indicate gleying.

Gleyed Humo-Ferric Podzol

Common horizon sequence: LFH or O, Aegj, <u>Bfgj</u>, BCg, Cg

These soils have the general properties specified for the Podzolic order and the Humo-Ferric Podzol great group. They differ from Orthic Humo-Ferric Podzols by having distinct or prominent mottles that indicate gleying, within 1 m of the surface. They usually have thick L, F, and H or O horizons and have neither ortstein, placic, duric, nor Bt horizons, a fragipan, nor an Ah horizon at least 10 cm thick.

Gleyed Ortstein Humo-Ferric Podzol

Common horizon sequence: LFH or O, Aegj, <u>Bfcgj</u>, Bfjcjgj, Cg

These soils have the general properties specified for the Podzolic order and the Humo-Ferric Podzol great group. They differ from Ortstein Humo-Ferric Podzols by having distinct or prominent mottles that indicate gleying, within 1 m of the surface. They usually have thick L, F, and H or O horizons and may have Ah, Ae, Bt, placic, or duric horizons, or a fragipan.

Gleyed Sombric Humo-Ferric Podzol

Common horizon sequence: LFH or O, <u>Ah</u>, Aegj, <u>Bfgj</u>, BCg, Cg

These soils have the general properties specified for the Podzolic order and the Humo-Ferric Podzol great group. They differ from Sombric Humo-Ferric Podzols by having distinct or prominent mottles that indicate gleying, within 1 m of the surface. They usually have thick L, F, and H or O horizons and may have an Ae horizon. They have neither ortstein, placic, duric, nor Bt horizons, nor a fragipan.

Chapter 11

Regosolic Order

Great Group	Subgroup
Regosol	Orthic Regosol O.R
	Cumulic Regosol CU.R
	Gleyed Regosol GL.R
	Gleyed Cumulic Regosol GLCU.R
Humic Regosol	Orthic Humic Regosol O.HR
	Cumulic Humic Regosol CU.HR
	Gleyed Humic Regosol GL.HR
	Gleyed Cumulic Humic Regosol GLCU.HR

A diagrammatic representation of profiles of some subgroups of the Regosolic order is shown in Figure 38. Individual subgroups may include soils that have horizon sequences different from those shown. In the description of each subgroup, presented later in this chapter, a common horizon sequence is given; diagnostic horizons are underlined and some other commonly occurring horizons are listed.

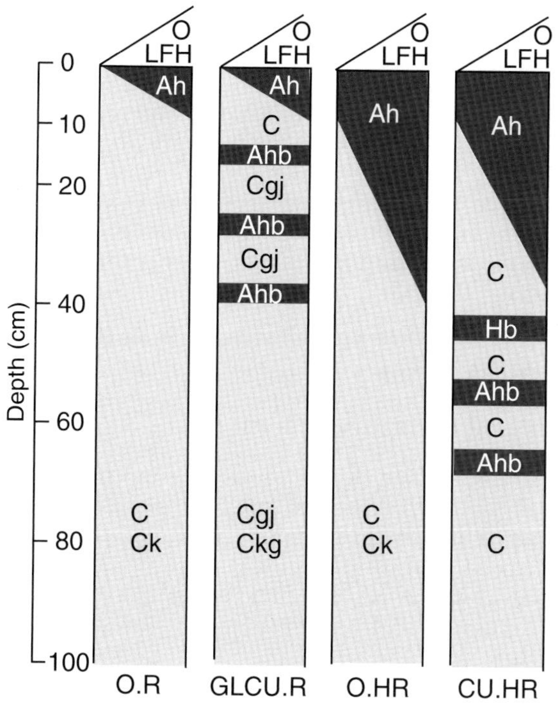

Regosolic soils do not contain a recognizable B horizon at least 5 cm thick and are therefore referred to as weakly developed. The lack of a developed pedogenic B horizon may result from any of a number of factors: youthfulness of the material, recent alluvium; instability of the material, colluvium on slopes subject to mass wasting; nature of the material, nearly pure quartz sand; climate, dry cold conditions. Regosolic soils are generally rapidly to imperfectly drained. They occur under a wide range of vegetation and climates.

Pedogenic development in Regosolic soils is too weak to form a recognizable B horizon that meets the requirements of any other order. They have none of the following: solonetzic B, Bt, podzolic B, Bm at least 5 cm thick, vertic horizon, evidence of gleying strong enough to meet the requirements of Gleysolic soils, organic surface horizons thick enough to meet the requirements of Organic soils, or permafrost within 1 or 2 m if the soils are strongly cryoturbated. They may have L, F, and H or O horizons. Also they may have an Ah or Ap horizon less than 10 cm thick or of any thickness if there is no underlying B horizon at least 5 cm thick and characteristics of the Ah or Ap do not satisfy the criteria of a chernozemic A horizon.

Distinguishing Regosolic Soils from Soils of Other Orders

Guidelines for distinguishing Regosolic soils from soils of other orders with which they might be confused follow:

Chernozemic Some Humic Regosols might be confused with Rego subgroups of

Figure 38 Diagrammatic horizon pattern of some subgroups of the Regosolic order.

117

Chernozemic soils. Such soils (usually Ah or Ap, C profiles) are classified as Humic Regosols if the Ah or Ap horizon fails to satisfy the requirements of a chernozemic A because of any of the following:

1. It is a moder rather than a mull A horizon.

2. It lacks structure.

3. The soil climate is outside the range specified for a chernozemic A.

4. It has a low base saturation.

Vertisolic Vertisolic soils must have both a slickenside and a vertic horizon within the control section. Regosolic soils do not have a vertic horizon.

Brunisolic Brunisolic soils must have a B horizon (Bm, Btj, Bfj) at least 5 cm thick. Regosolic soils usually do not have a B horizon and in those that do it is less than 5 cm thick.

Gleysolic Some Gleysolic soils resemble Regosolic soils by having very weakly developed horizons. However, Gleysolic soils must have dull colors or mottles that indicate strong gleying; Regosolic soils do not.

Cryosolic Cryosolic soils have permafrost within 1 m of the surface or 2 m if strongly cryoturbated; Regosolic soils do not.

The Regosolic order is divided into two great groups based on the presence or absence of a significantly developed Ah or dark colored Ap horizon as shown in the Regosolic order chart.

Subgroups are based upon evidence of either relative stability of the material or periodic deposition of material and of gleying. Regosols with saline, cryic, and lithic features are differentiated taxonomically either at the family (lithic, some cryic) or the series (saline) levels. These features may also be indicated as phases of a subgroup, great group, or order.

Regosol

These Regosolic soils do not have an Ah or dark-colored Ap horizon at least 10 cm thick at the mineral soil surface. They may have buried mineral–organic layers and organic surface horizons, but no B horizon at least 5 cm thick.

Orthic Regosol

Common horizon sequence: C

These soils have the properties specified for the Regosolic order and the Regosol great group. They are identified by the following properties:

1. If they have an A horizon, it is less than 10 cm thick.

2. The B horizon is less than 5 cm thick or absent.

3. They have a small amount of organic matter in the control section, which is indicated by a uniform color so that the color value difference between layers is less than one Munsell unit.

4. They are well drained and lack any evidence of gleying within the upper 50 cm.

Cumulic Regosol

Common horizon sequence: C, Ahb, C

These soils have the properties specified for the Regosolic order and the Regosol great group. They differ from Orthic Regosols by having below the surface, or below any thin Ah horizon, layers that vary in color value by 1 or more units, or organic matter contents that vary irregularly with depth. They lack evidence of gleying within 50 cm of the mineral surface. Commonly these soils result from material deposited during intermittent floods.

Gleyed Regosol

Common horizon sequence: Cgj

These soils have the properties specified for the Regosolic order and the Regosol great group. They differ from Orthic Regosols by having faint to distinct mottles that indicate gleying within 50 cm of the mineral surface.

Regosolic Order

	Regosol	Humic Regosol
Ah or dark colored Ap	<10 cm	≥ 10 cm thick
Bm	absent or <5 cm	absent or <5 cm

Gleyed Cumulic Regosol

Common horizon sequence: Cgj, Ahb, Cgj

These soils have the properties specified for the Regosolic order and the Regosol great group. They differ from Cumulic Regosols by having faint to distinct mottles that indicate gleying within 50 cm of the mineral surface.

Humic Regosol

These Regosolic soils have an Ah or dark colored Ap horizon at least 10 cm thick at the mineral surface. They may have organic surface horizons and buried mineral–organic horizons. They do not have a B horizon at least 5 cm thick.

Orthic Humic Regosol

Common horizon sequence: Ah, C

These soils have the properties specified for the Regosolic order and the Humic Regosol great group. They are identified by the following properties:

1. They have an Ah or dark-colored Ap horizon at least 10 cm thick.
2. They have no B horizon or the B horizon is less than 5 cm thick.
3. They have a low content of organic matter throughout the control section below the A horizon; this is reflected in a uniform color with differences of Munsell color value of less than one unit between layers.
4. They lack faint to distinct mottling that indicates gleying within the upper 50 cm.

Cumulic Humic Regosol

Common horizon sequence: Ah, C, Ahb, C

These soils have the properties specified for the Regosolic order and the Humic Regosol great group. They differ from Orthic Humic Regosols by having layers below the Ah or Ap horizon that vary in color value by one or more units, or organic matter contents that vary irregularly with depth. They do not have faint to distinct mottles that indicate gleying within 50 cm of the mineral surface. Commonly these soils result from either mass wasting of soil downslope or intermittent flooding and deposition of material.

Gleyed Humic Regosol

Common horizon sequence: Ah, Cgj

These soils have the properties specified for the Regosolic order and the Humic Regosol great group. They differ from Orthic Humic Regosols by having faint to distinct mottles that indicate gleying within 50 cm of the mineral surface.

Gleyed Cumulic Humic Regosol

Common horizon sequence: Ah, Cgj, Ahb, Cgj

These soils have the properties specified for the Regosolic order and the Humic Regosol great group. They differ from Cumulic Humic Regosols by having faint to distinct mottles that indicate gleying within 50 cm of the mineral surface.

Chapter 12

Solonetzic Order

Great Group	Subgroup
Solonetz	Brown Solonetz B.SZ
	Dark Brown Solonetz DB. SZ
	Black Solonetz BL.SZ
	Alkaline Solonetz A.SZ
	Gleyed Brown Solonetz GLB.SZ
	Gleyed Dark Brown Solonetz GLDB.SZ
	Gleyed Black Solonetz GLBL.SZ
Solodized Solonetz	Brown Solodized Solonetz B.SS
	Dark Brown Solodized Solonetz DB.SS
	Black Solodized Solonetz BL.SS
	Dark Gray Solodized Solonetz DG.SS
	Gray Solodized Solonetz G.SS
	Gleyed Brown Solodized Solonetz GLB.SS
	Gleyed Dark Brown Solodized Solonetz GLDB.SS
	Gleyed Black Solodized Solonetz GLBL.SS
	Gleyed Dark Gray Solodized Solonetz GLDG.SS
	Gleyed Gray Solodized Solonetz GLG.SS
Solod	Brown Solod B.SO
	Dark Brown Solod DB.SO
	Black Solod BL.SO
	Dark Gray Solod DG.SO
	Gray Solod G.SO
	Gleyed Brown Solod GLB.SO
	Gleyed Dark Brown Solod GLDB.SO
	Gleyed Black Solod GLBL.SO
	Gleyed Dark Gray Solod GLDG.SO
	Gleyed Gray Solod GLG.SO
Vertic Solonetz	Brown Vertic Solonetz BV.SZ
	Dark Brown Vertic Solonetz DBV.SZ
	Black Vertic Solonetz BLV.SZ
	Gleyed Brown Vertic Solonetz GLBV.SZ
	Gleyed Dark Brown Vertic Solonetz GLDBV.SZ
	Gleyed Black Vertic Solonetz GLBLV.SZ

A diagrammatic representation of profiles of some subgroups of the Solonetzic order is shown in Figure 39. Individual subgroups may include soils that have horizon sequences different from those shown. In the description of each subgroup, presented later in this chapter, a common horizon sequence is given; diagnostic horizons are underlined and some other commonly occurring horizons are listed.

Soils of the Solonetzic order have B horizons that are very hard when dry and swell to a sticky mass of very low permeability when wet. Typically the solonetzic B horizon has prismatic or columnar macrostructure that breaks to hard to extremely hard (when dry) blocky peds with dark coatings. They occur on saline parent materials in some areas of the semiarid to subhumid Interior Plains in association with Chernozemic soils and to a lesser extent with Luvisolic and Gleysolic soils. Most Solonetzic soils are associated with a vegetative cover of grasses and forbs. Although some occur under tree cover, it is thought that the trees did not become established until solodization was well under way.

Solonetzic soils are thought to have developed from parent materials that were more or less uniformly salinized with salts high in sodium. Leaching of salts by descending rainwater presumably results in deflocculation of the sodium-saturated colloids. The peptized colloids are apparently

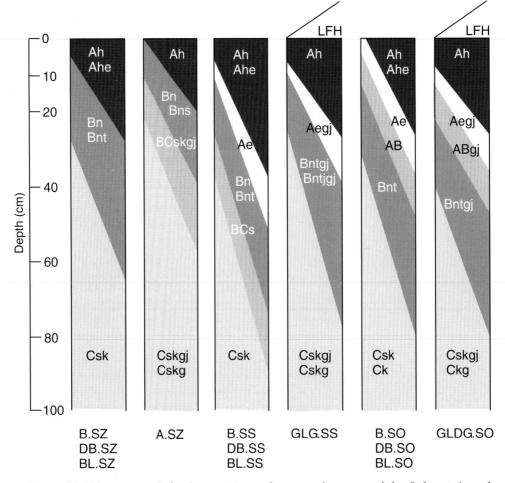

Figure 39 Diagrammatic horizon pattern of some subgroups of the Solonetzic order.

carried downward and deposited in the B horizon. Further leaching results in depletion of alkali cations in the A horizon, which becomes acidic and a platy Ae horizon usually develops. The underlying B horizon usually consists of darkly stained, fused, intact columnar peds. Structural breakdown of the upper part of the B horizon apparently occurs at an advanced stage of development as exchangeable sodium is leached downward. At this stage the solonetzic B usually breaks readily to blocky peds coated with white silicious powder. Complete destruction of the solonetzic B horizon is the most advanced stage of solodization. The rate of evolution through the stages of development depends on the salt content and hydraulic conductivity of the parent material and on the climate.

Most Solonetzic soils in Canada have a neutral to acidic A horizon indicating that some solodization has occurred. Soils with strongly alkaline A horizons, an early stage of Solonetzic soil formation, are uncommon in

Canada. As solodization proceeds, the horizons of salt and lime accumulation move downward from the B to the C horizon. In most Solonetzic soils the saturation extract of the C horizon has a conductivity of at least 4 mS cm^{-1}. Solodization is arrested where saline groundwater is within capillary reach of the solum, and resalinization may occur in groundwater discharge areas.

Soils of the Solonetzic order have a solonetzic B horizon (Bn or Bnt). This horizon has columnar or prismatic structure, is hard to extremely hard when dry, and has a ratio of exchangeable Ca to Na of 10 or less. The macrostructural units usually break to form hard to extremely hard blocky peds with dark coatings. Solonetzic soils do not have permafrost within 1 m of the surface, a surface organic layer 60 cm or more in thickness if fibric or 40 cm or more if mesic or humic, a podzolic B horizon, or evidence of gleying strong enough to meet the criteria for Gleysolic soils.

Distinguishing Solonetzic Soils from Soils of Other Orders

Chernozemic   Soils having a chernozemic A horizon and a solonetzic B horizon are classified as Solonetzic. However, sometimes the structure and consistence of the Bnt horizons of Solonetzic soils are similar to the Btnj horizons of some Solonetzic subgroups of Chernozemic soils. In borderline cases the ratio of exchangeable Ca to Na determines the classification.

Luvisolic   Some Luvisolic soils are similar to Gray and Dark Gray Solods. The soils having a Bnt horizon rather than a Btnj are classified as Solonetzic.

Gleysolic   Some soils have solonetzic B horizons, dull colors, and mottling that indicates strong gleying, within 50 cm of the mineral soil surface. These soils are classified as Gleysolic.

Vertisolic   Some Solonetzic soils have a slickenside horizon but Vertisolic soils have both a vertic and a slickenside horizon. Solonetzic soils do not.

The Solonetzic order includes four great groups: Solonetz, Solodized Solonetz, Solod, and Vertic Solonetz. They are separated based on the degree of expression of the Ae horizon, the breakdown of the upper part of the B horizon, and the occurrence of Vertic features as indicated in the Solonetzic order chart.

Subgroups are separated on the basis of features indicating different climatic zones as reflected in the color of the A horizon and on the presence or absence of gleying. Lithic features are now recognized taxonomically at the family level or as a phase of subgroups, great groups, or orders.

Solonetz

The soils of this great group usually have an Ah, Ahe, or Ap horizon overlying the solonetzic B horizon; any Ae horizon is thin and usually discontinuous. The boundary between the A and B horizons is abrupt and usually occurs within 20 cm of the surface. The solonetzic B horizon (Bn or Bnt) is hard and commonly massive, breaking to angular blocky structure. It has a low hydraulic conductivity. Dark stainings occur commonly on ped surfaces in the B horizon, which usually has a neutral to strongly alkaline reaction and may contain carbonates. The C horizon is saline and usually calcareous.

Soils of the Solonetzic great group occur throughout the area of Solonetzic soils. The great group is usually associated with parent materials of high salinity on relatively arid sites.

These soils have a solonetzic B horizon (Bn or Bnt) and lack a continuous Ae horizon at least 2 cm thick. The Solonetz great group is divided into seven subgroups.

Brown Solonetz

Common horizon sequence: Ah, Bn or Bnt, Csk

Soils of this subgroup have a solonetzic B horizon as defined for the Solonetzic order and the properties specified for the Solonetz great group. They are associated with grass and forb vegetation and a subarid to semiarid climate. Areas of these soils often have patchy microrelief caused by differential erosion. The B horizon is exposed in some eroded pits. Plant growth in the eroded pits is usually very sparse.

Brown Solonetzs have either an Ah, Ahe, or Ap horizon with color values higher than 4.5 dry and chromas usually higher than 1.5 dry, or an exposed solonetzic B horizon. They have neither a strongly alkaline (pH of 8.5 or more) A horizon nor faint to distinct mottles that indicate gleying within the upper 50 cm.

Solonetzic Order

	Solonetz	Solodized Solonetz	Solod	Vertic Solonetz
Ae horizon	no continuous Ae ≥ 2 cm thick	Ae ≥ 2 cm thick	Ae ≥ 2 cm thick	any features of Solonetz, Solodized Solonetz, or Solod
B horizon		intact, columnar Bnt or Bn	distinct AB or BA (disintegrating Bnt)	
Vertic features				slickenside

Dark Brown Solonetz

Common horizon sequence: <u>Ah</u>, <u>Bn</u> or <u>Bnt</u>, Csk

Soils of this subgroup have a solonetzic B horizon as defined for the Solonetzic order and the properties specified for the Solonetz great group. They are associated with mesophytic grasses and forbs in a semiarid climate. Areas of Dark Brown Solonetzs often have patchy microrelief caused by differential erosion of the A horizon. However, bare eroded pits are not as common as in areas of Brown Solonetzs.

Dark Brown Solonetzs have either an Ah, Ahe, or Ap horizon with color values lower than 3.5 moist and 3.5–4.5 dry, or an exposed solonetzic B horizon. Chromas of the surface horizon are usually higher than 1.5 dry. The soils have neither strongly alkaline (pH of 8.5 or more) A horizons nor faint to distinct mottles that indicate gleying within the upper 50 cm.

Black Solonetz

Common horizon sequence: <u>Ah</u>, <u>Bnt</u>, Csk

Soils of this subgroup have a solonetzic B horizon as defined for the Solonetzic order and the properties specified for the Solonetz great group. They are associated mainly with the growth of mesophytic grasses and forbs in a subhumid climate. However, they occur also in areas of discontinuous shrub and tree cover with a ground cover of forbs and grasses. The microrelief associated with differential erosion of the A horizon in areas of Brown and Dark Brown Solonetzs occurs rarely in areas of Black Solonetzs.

Black Solonetzs have an Ah, Ahe, or Ap horizon with color values lower than 3.5 moist and dry, and chromas usually less than 2 moist and dry. They have neither a strongly alkaline (pH of 8.5 or more) A horizon nor faint to distinct mottles that indicate gleying within the upper 50 cm.

Alkaline Solonetz

Common horizon sequence: <u>Ah</u>, <u>Bn</u>, Csk

Soils of this subgroup have a solonetzic B horizon as defined for the Solonetzic order and the properties specified for the Solonetz great group. They are associated with highly saline materials and with mesophytic grasses and forbs that include alkali-tolerant species. The alkaline A horizon may be of any color including brown, black, and gray. Alkaline Solonetzs occupy a minor area and commonly occur in groundwater discharge sites.

Alkaline Solonetzs have a strongly alkaline A horizon (pH of 8.5 or more) and a solonetzic B horizon. They may have faint to distinct mottles that indicate gleying within the upper 50 cm.

Gleyed Brown Solonetz

Common horizon sequence: <u>Ah</u>, <u>Bngj</u>, Cskgj

Soils of this subgroup have a solonetzic B horizon as defined for the Solonetzic order and the properties specified for the Solonetz great group. They differ from Brown Solonetzs by having faint to distinct mottles that indicate gleying within 50 cm of the mineral surface.

Gleyed Dark Brown Solonetz

Common horizon sequence: <u>Ah</u>, <u>Bngj</u>, Cskgj

Soils of this subgroup have a solonetzic B horizon as defined for the Solonetzic order and the properties specified for the Solonetz great group. They differ from Dark Brown Solonetzs by having faint to distinct mottles that indicate gleying within 50 cm of the mineral surface.

Gleyed Black Solonetz

Common horizon sequence: <u>Ah</u>, <u>Bntgj</u>, Cskgj

Soils of this subgroup have a solonetzic B horizon as defined for the Solonetzic order and the properties specified for the Solonetz great group. They differ from Black Solonetzs by having faint to distinct mottles that indicate gleying within 50 cm of the mineral surface.

Solodized Solonetz

The soils of this great group have a distinct Ae horizon and a hard columnar or prismatic solonetzic B horizon. The Ae horizon is at least 2 cm thick, usually has well-developed platy structure, and its reaction is acid to neutral. It has an abrupt boundary to a strongly expressed, intact, solonetzic B horizon. The Bnt or Bn horizon is hard to extremely hard when dry and usually has a columnar macrostructure with white-capped, coherent columns that break to blocky peds with dark coatings. Usually the Bnt horizon has many clay skins. The C horizon is saline and commonly calcareous.

The Solodized Solonetz great group is divided into eight subgroups.

Brown Solodized Solonetz

Common horizon sequence: <u>Ah, Ae, Bn</u> or <u>Bnt,</u> Csk

Soils of this subgroup have a solonetzic B horizon as defined for the Solonetzic order and the properties specified for the Solodized Solonetz great group. They are associated with grass and forb vegetation and a subarid to semiarid climate. Areas of this subgroup often have irregular microrelief caused by differential erosion of the A horizons. The eroded pits usually support only a very sparse plant cover.

Brown Solodized Solonetzs have an Ah, Ahe, or Ap horizon with color values higher than 4.5 dry and chromas usually higher than 1.5 dry. They have Ae and solonetzic B horizons as specified for the great group. They do not have faint to distinct mottles that indicate gleying within the upper 50 cm.

Dark Brown Solodized Solonetz

Common horizon sequence: <u>Ah, Ae, Bn</u> or <u>Bnt,</u> Csk

Soils of this subgroup have a solonetzic B horizon as defined for the Solonetzic order and the properties specified for the Solodized Solonetz great group. They are associated with mesophytic grasses and forbs in a semiarid climate. Areas of Dark Brown Solodized Solonetzs commonly have irregular microrelief caused by differential erosion of the A horizons.

Dark Brown Solodized Solonetzs have an Ah, Ahe, or Ap horizon with color values lower than 3.5 moist and 3.5–4.5 dry and chromas usually higher than 1.5 dry. They have Ae and solonetzic B horizons as specified for the great group. They do not have faint to distinct mottles that indicate gleying within the upper 50 cm.

Black Solodized Solonetz

Common horizon sequence: <u>Ah, Ae, Bnt</u> or <u>Bn,</u> Csk

Soils of this subgroup have a solonetzic B horizon as defined for the Solonetzic order and the properties specified for the Solodized Solonetz great group. They are associated mainly with mesophytic grass and forb vegetation and a subhumid climate. However, they also occur in areas of discontinuous shrub and tree vegetation with a ground cover of forbs and grasses. Differential erosion of the A horizon, which is common in areas of Brown and Dark Brown Solodized Solonetzs, seldom occurs in these soils.

Black Solodized Solonetzs have an Ah, Ahe, or Ap horizon with color values lower than 3.5 dry and chromas usually lower than 2 dry. They have Ae and solonetzic B horizons as specified for the great group. They do not have faint to distinct mottles that indicate gleying within the upper 50 cm.

Dark Gray Solodized Solonetz

Common horizon sequence: <u>Ahe, Ae, Bnt,</u> Csk

Soils of this subgroup have a solonetzic B horizon as defined for the Solonetzic order and the properties specified for the Solodized Solonetz great group. They are associated mainly with sparse forest vegetation, a ground cover of forbs and grasses, a subhumid climate, and saline parent material.

Dark Gray Solodized Solonetzs have an Ahe or Ap horizon with color values of 3.5–4.5 dry and chromas usually lower than 2 moist or dry. They have Ae and solonetzic B horizons as specified for the great group. They lack faint to distinct mottles that indicate gleying within the upper 50 cm.

Gray Solodized Solonetz

Common horizon sequence: <u>Ahe, Ae, Bnt,</u> Csk

Soils of this subgroup have a solonetzic B horizon as defined for the Solonetzic order and the properties specified for the Solodized Solonetz great group. They are usually associated with sparse forest vegetation, a subhumid climate, and Eluviated Black Chernozems or Dark Gray Luvisols developed from parent materials of lower salinity.

Gray Solodized Solonetzs have an Ahe, Ae, or Ap horizon with color values higher than 4.5 dry and chromas usually lower than 2 moist or dry. They have Ae and solonetzic B horizons as specified for the great group. They do not have faint to distinct mottles that indicate gleying within the upper 50 cm.

Gleyed Brown Solodized Solonetz

Common horizon sequence: <u>Ah, Aegj, Bngj</u>, Cskgj

Soils of this subgroup have a solonetzic B horizon as defined for the Solonetzic order and the properties specified for the Solodized Solonetz great group. They differ from Brown Solodized Solonetzs by having faint to distinct mottles that indicate gleying within 50 cm of the mineral surface.

Gleyed Dark Brown Solodized Solonetz

Common horizon sequence: <u>Ah, Aegj, Bngj</u> or <u>Bntgj</u>, Cskgj

Soils of this subgroup have a solonetzic B horizon as defined for the Solonetzic order and the properties specified for the Solodized Solonetz great group. They differ from Dark Brown Solodized Solonetzs by having faint to distinct mottles that indicate gleying within 50 cm of the mineral surface.

Gleyed Black Solodized Solonetz

Common horizon sequence: <u>Ah, Aeg, Bntgj</u>, Cskgj

Soils of this subgroup have a solonetzic B horizon as defined for the Solonetzic order and the properties specified for the Solodized Solonetz great group. They differ from Black Solodized Solonetzs by having faint to distinct mottling that indicates gleying within 50 cm of the mineral surface.

Gleyed Dark Gray Solodized Solonetz

Common horizon sequence: <u>Ahe, Aegj, Bntgj</u>, Cskg

Soils of this subgroup have a solonetzic B horizon as defined for the Solonetzic order and the properties specified for the Solodized Solonetz great group. They differ from Dark Gray Solodized Solonetzs by having faint to distinct mottles that indicate gleying within 50 cm of the mineral surface.

Gleyed Gray Solodized Solonetz

Common horizon sequence: <u>Ahe, Aegj, Bntgj</u>, Cskg

Soils of this subgroup have a solonetzic B horizon as defined for the Solonetzic order and the properties specified for the Solodized Solonetz great group. They differ from Gray Solodized Solonetzs by having faint to distinct mottles that indicate gleying within 50 cm of the mineral surface.

Solod

Besides an Ae horizon, the soils of this great group have an AB or BA horizon, which appears to have developed through degradation of the upper part of a former solonetzic B horizon. Both the Ae and AB horizons are usually platy. The upper boundary of the solonetzic B horizon is gradual and usually at greater depth than the B horizon of the associated Solonetzs and Solodized Solonetzs. The solonetzic B horizon commonly does not have strong columnar structure and usually breaks to blocky peds that are hard to very hard when dry. Clay skins are common to frequent in the B horizon. The Cs or Csa horizon commonly occurs below a Cca or Ck horizon.

Solods have a distinct Ae horizon at least 2 cm thick, an AB or BA horizon at least 5 cm thick, and a solonetzic B horizon. The great group is divided into eight subgroups.

Brown Solod

Common horizon sequence: <u>Ah, Ae, AB, Bnt</u>, Ck, Csk

Soils of this subgroup have a solonetzic B horizon as defined for the Solonetzic order and the properties specified for the Solod great group. They are associated with grass and forb vegetation and a subarid to semiarid climate. In areas of these soils there is commonly evidence of previously eroded pits, but they are usually shallow and grass covered.

Brown Solods have an Ah, Ahe, or Ap horizon with color values higher than 4.5 dry and chromas usually higher than 1.5 dry. They have Ae, AB or BA, and solonetzic B horizons as specified for the great group. They do not have faint to distinct mottles that indicate gleying within the upper 50 cm.

Dark Brown Solod

Common horizon sequence: <u>Ahe, Ae, AB, Bnt</u>, Ck, Csk

Soils of this subgroup have a solonetzic B horizon as defined for the Solonetzic order and the properties specified for the Solod great group. They are associated with mesophytic grasses and forbs in a semiarid to subhumid climate. Evidence of former eroded pits is common in areas of these soils, but the depressions are shallow and grass covered.

Dark Brown Solods have an Ah, Ahe or Ap horizon with color values lower than 3.5 moist and 3.5–4.5 dry and chromas usually higher than 1.5 dry. They have Ae, AB or BA, and solonetzic B horizons as specified for the great group. They do not have faint to distinct mottles that indicate gleying within the upper 50 cm.

Black Solod

Common horizon sequence: <u>Ahe</u>, <u>Ae</u>, <u>AB</u>, <u>Bnt</u>, Ck, Csk

Soils of this subgroup have a solonetzic B horizon as defined for the Solonetzic order and the properties specified for the Solod great group. They are associated mainly with mesophytic grasses and forbs in a subhumid climate, but they also occur in areas of thin or discontinuous shrub and tree cover.

Black Solods have an Ah, Ahe, or Ap horizon with color values lower than 3.5 dry and chromas usually higher than 1.5 dry. They have Ae, AB or BA, and solonetzic B horizons as specified for the great group. They do not have faint to distinct mottles that indicate gleying within the upper 50 cm.

Dark Gray Solod

Common horizon sequence: <u>Ahe</u>, <u>Ae</u>, <u>AB</u>, <u>Bnt</u>, Ck, Csk

Soils of this subgroup have a solonetzic B horizon as defined for the Solonetzic order and the properties specified for the Solod great group. They are associated with forest, shrub, and grass vegetation, a subhumid climate, and Eluviated Black Chernozems or Dark Gray Luvisols developed in materials of lower salinity.

Dark Gray Solods have an Ah, Ahe or Ap horizon with color values of 3.5–4.5 dry and chromas usually lower than 2 moist or dry. They have Ae, AB, and solonetzic B horizons as specified for the great group. They do not have faint to distinct mottles that indicate gleying within the upper 50 cm.

Gray Solod

Common horizon sequence: <u>Ahe</u>, <u>Ae</u>, <u>AB</u>, <u>Bnt</u>, Ck, Csk

Soils of this subgroup have a solonetzic B horizon as defined for the Solonetzic order and the properties specified for the Solod great group. They are associated with forest vegetation, a subhumid climate, and Dark Gray Luvisols developed in materials of lower salinity.

Gray Solods have an Ahe, Ae or Ap horizon with color values higher than 4.5 dry and chromas usually lower than 2 dry. They have Ae, AB or BA, and solonetzic B horizons as specified for the great group. They do not have faint to distinct mottles that indicate gleying within the upper 50 cm.

Gleyed Brown Solod

Common horizon sequence: <u>Ah</u>, <u>Aegj</u>, <u>ABgj</u>, <u>Bntgj</u>, Cskg

Soils of this subgroup have a solonetzic B horizon as defined for the Solonetzic order and the properties specified for the Solod great group. They differ from Brown Solods by having faint to distinct mottles that indicate gleying within 50 cm of the mineral surface.

Gleyed Dark Brown Solod

Common horizon sequence: <u>Ah</u>, <u>Aegj</u>, <u>ABgj</u>, <u>Bntgj</u>, Cskg

Soils of this subgroup have a solonetzic B horizon as defined for the Solonetzic order and the properties specified for the Solod great group. They differ from Dark Brown Solods by having faint to distinct mottles that indicate gleying within 50 cm of the mineral surface.

Gleyed Black Solod

Common horizon sequence: <u>Ah</u>, <u>Aegj</u>, <u>ABgj</u>, <u>Bntgj</u>, Cskg

Soils of this subgroup have a solonetzic B horizon as defined for the Solonetzic order and the properties specified for the Solod great group. They differ from Black Solods by having faint to distinct mottles that indicate gleying within 50 cm of the mineral surface.

Gleyed Dark Gray Solod

Common horizon sequence: <u>Ahe</u>, <u>Aegj</u>, <u>ABgj</u>, <u>Bntgj</u>, Cskg

Soils of this subgroup have a solonetzic B horizon as defined for the Solonetzic order and the properties specified for the Solod great group. They differ from Dark Gray Solods by having faint to distinct mottles that indicate gleying within 50 cm of the mineral surface.

Gleyed Gray Solod

Common horizon sequence: <u>Ahe</u>, <u>Aegj</u>, <u>ABgj</u>, <u>Bntgj</u>, Cskg

Soils of this subgroup have a solonetzic B horizon as defined for the Solonetzic order and the properties specified for the Solod great group. They differ from Gray Solods by having faint to distinct mottles that indicate gleying within 50 cm of the mineral surface.

Vertic Solonetz

The soils of this great group have horizons that are characteristic of any of the other three great groups of Solonetzic soils. In addition, they have properties that indicate a intergrading to the Vertisolic order. Specifically they have a slickenside horizon, the upper boundary of which occurs within 1 m of the mineral surface. They may also have a weak vertic horizon.

Brown Vertic Solonetz

Common horizon sequence: <u>Ah</u>, Ae or AB, <u>Bn</u> or <u>Bnt</u>, Bnvj or Bntvj, <u>Bnss</u> or <u>Bntss</u> or <u>Cskss</u>, Csk

Soils of this subgroup have a solonetzic B horizon as defined for the Solonetzic order and the properties specified for the Vertic Solonetz great group. Brown Vertic Solonetzs have either an Ah, Ahe, or Ap horizon with color values higher than 4.5 dry and chromas usually higher than 1.5 dry. They also have properties that indicate intergrading to the Vertisolic order. Specifically, they have a slickenside horizon (Bnss or Cskss), the upper boundary of which occurs within 1 m of the mineral surface. They may also have a weak vertic horizon (Bntvj).

Dark Brown Vertic Solonetz

Common horizon sequence: <u>Ah</u>, Ae or AB, <u>Bn</u> or <u>Bnt</u>, Bnvj or Bntvj, <u>Bnss</u> or <u>Bntss</u> or <u>Cskss</u>, Csk

Soils of this subgroup have a solonetzic B horizon as defined for the Solonetzic order and the properties specified for the Vertic Solonetz great group. Dark Brown Vertic Solonetzs have either an Ah, Ahe, or Ap horizon with color values lower than 3.5 moist and between 3.5–4.5 dry. They also have properties that indicate intergrading to the Vertisolic order. Specifically, they have a slickenside horizon (Bnss or Cskss), the upper boundary of which occurs within 1 m of the mineral surface. They may also have a weak vertic horizon (Bntvj).

Black Vertic Solonetz

Common horizon sequence: <u>Ah</u>, Ae or AB, <u>Bn</u> or <u>Bnt</u>, Bnvj or Bntvj, <u>Bnss</u> or <u>Bntss</u> or <u>Cskss</u>, Csk

Soils of this subgroup have a solonetzic B horizon as defined for the Solonetzic order and the properties specified for the Vertic Solonetz great group. Black Vertic Solonetzs have either an Ah, Ahe, or Ap horizon with color values lower than 3.5 moist and dry and chromas usually less than 2. Also they have properties that indicate inter- grading to the Vertisolic order. Specifically, they have a slickenside horizon (Bnss or Cskss), the upper boundary of which occurs within 1 m of the mineral surface. They may also have a weak vertic horizon (Bntvj).

Gleyed Brown Vertic Solonetz

Common horizon sequence: <u>Ah</u>, Ae or AB, <u>Bngj</u> or <u>Bntgj</u>, Bngjvj or Bntgjvj, <u>Bngjss</u> or <u>Bntgjss</u> or <u>Cskgjss</u>, Cskgj or Cskg

Soils of this subgroup have a solonetzic B horizon as defined for the Solonetzic order and the properties specified for the Vertic Solonetz great group. Gleyed Brown Vertic Solonetzs have either an Ah, Ahe, or Ap horizon with color values higher than 4.5 dry and chromas usually higher than 1.5 dry. They also have properties that indicate intergrading to the Vertisolic order. Specifically, they have a slickenside horizon (Bngjss, Bntgjss or Cskgjss), the upper boundary of which occurs within 1 m of the mineral surface. They may also have a weak vertic horizon (Bngjvj or Bntgjvj). They differ from Brown Vertic Solonetzs by having faint to distinct mottles that indicate gleying within 50 cm of the mineral surface.

Gleyed Dark Brown Vertic Solonetz

Common horizon sequence: <u>Ah</u>, Ae or AB, <u>Bngj</u> or <u>Bntgj</u>, Bngjvj or Bntgjvj, <u>Bngjss</u>, <u>Bntgjss</u> or <u>Cskgjss</u>, Cskgj or Cskg

Soils of this subgroup have a solonetzic B horizon as defined for the Solonetzic order and the properties specified for the Vertic Solonetz great group. Gleyed Dark Brown Vertic Solonetzs have either an Ah, Ahe, or Ap horizon with color values lower than 3.5 moist and between 3.5–4.5 dry. They also have properties that indicate intergrading to the Vertisolic order. Specifically, they have a slickenside horizon (Bngjss, Bntgjss or Cskgjss), the upper boundary of which occurs within 1 m of the mineral surface. They may also have a weak vertic horizon (Bngjvj or Bntgjvj). They differ from Dark Brown Vertic Solonetzs by having faint to distinct mottles that indicate gleying within 50 cm of the mineral surface.

Gleyed Black Vertic Solonetz

Common horizon sequence: <u>Ah</u>, Ae or AB, <u>Bngj</u> or <u>Bntgj</u>, Bngjvj or Bntgjvj, <u>Bngjss</u> or <u>Bntgjss</u> or <u>Cskgjss</u>, Cskgj or Cskg

Soils of this subgroup have a solonetzic B horizon as defined for the Solonetzic order and the properties specified for the Vertic Solonetz great group. Gleyed Black Vertic Solonetzs have either an Ah, Ahe, or Ap horizon with color values lower than 3.5 moist and dry and chromas usually less than 2. They also have properties that indicate intergrading to the Vertisolic order. Specifically, they have a slickenside horizon (Bngjss, Bntgjss or Cskgjss), the upper boundary of which occurs within 1 m of the mineral surface. They may also have a weak vertic horizon (Bngjvj or Bntgjvj). They differ from Black Vertic Solonetzs by having faint to distinct mottles that indicate gleying within 50 cm of the mineral surface.

Chapter 13

Vertisolic Order

Great Group	Subgroup
Vertisol	Orthic Vertisol O.V
	Gleyed Vertisol GL.V
	Gleysolic Vertisol GLC.V
Humic Vertisol	Orthic Humic Vertisol O.HV
	Gleyed Humic Vertisol GL.HV
	Gleysolic Humic Vertisol GLC.HV

A diagrammatic representation of profiles of some subgroups of the Vertisolic order is given in Figure 40. The subgroups include soils that may have horizon sequences different from those indicated. In the description of each subgroup, presented later in this chapter, a common horizon sequence is given; diagnostic horizons are underlined and some other commonly occurring horizons are listed.

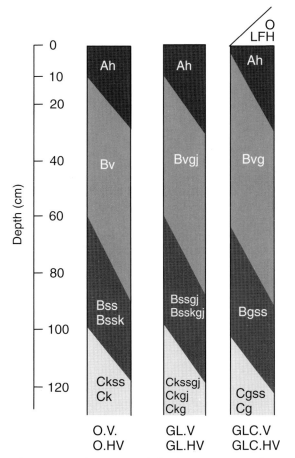

Figure 40 Diagrammatic horizon pattern of some subgroups of the Vertisolic order.

Soils of the Vertisolic order occur in heavy textured materials (≥60% clay of which at least half is smectite) and have shrink–swell characteristics that are diagnostic of Vertisolic soils. However, they lack the degree or kind of horizon development diagnostic of soils of other orders, except as noted below. The central concept of the order is that of soils in which cracking, argillipedoturbation, and mass movement, as evidenced by the presence of slickensides and severe disruption within the control section, are the dominant soil-forming processes. These disturbances, caused by shrinking and swelling, the result of wetting and drying cycles, are strong enough that horizons diagnostic of other soil orders have either been prevented from forming or have been severely disrupted. However, in terms of color and organic matter content, they may have a Chernozemic-like A horizon at the surface, or they may have features diagnostic of the Gleysolic order. The major areas of Vertisolic soils occur in the cool, subarid to subhumid, grassland portion of the Interior Plains of western Canada. Minor areas of Vertisolic soils may occur in valleys of the South Cordilleran Region, in parts of the South Boreal Region, and in the Cool Temperate Regions of central Canada. (Brierley et al. 1996)

Soils of the Vertisolic order are defined based on the occurrence of BOTH a slickenside horizon and a vertic horizon within the control section. A slickenside horizon (ss) is one in which several (more than two) slickensides, which may be intersecting, can be recognized. A vertic horizon (v) is one that has been strongly affected by argillipedo-turbation, as a result of shrinking and swelling, and is characterized by the presence of:

1. Irregular shaped, randomly oriented, intrusions of displaced materials within the solum.
2. Vertical cracks, often containing sloughed-in surficial materials.

The disruption within this horizon, resulting from the shrinking and swelling process, is either strong enough to prevent horizons diagnostic of other orders from forming or, if diagnostic horizons are present, they are disrupted to the extent that they are no longer continuous throughout the pedon and their orientation has been severely changed.

Distinguishing Vertisolic Soils from Soils of Other Orders

Vertisolic soils are distinguished from soils of other orders by having both a vertic and slickenside horizon. They also have no undisturbed horizons that are diagnostic of other orders, except that they may have a Chernozemic-like A horizon at the surface or may have features diagnostic of the Gleysolic order, or both.

Guidelines for distinguishing Vertisolic soils from soils of the Chernozemic and Gleysolic orders follow.

Chernozemic Vertisolic soils are distinguished from soils of the Chernozemic order based on the following:

1. The surface horizon (Ah), when dry, has massive structure and is hard.
2. No other undisturbed A and B horizons common to Chernozemic soils are present, other than the Ap or Ah horizon.
3. Both a slickenside horizon (ss), the upper boundary of which occurs within 1 m of the mineral surface, and a vertic horizon (v) are present. The occurrence of these horizons together takes precedence over

the occurrence of a Chernozemic-like A horizon.

Gleysolic Vertisolic soils are distinguished from Gleysolic soils on the basis that both a slickenside horizon (ss), the upper boundary of which occurs within 1 m of the mineral surface, and a vertic horizon (v) are present in Vertisolic soils. The occurrence of these horizons together takes precedence over the occurrence of gley features diagnostic of the Gleysolic order.

Vertisolic soils are divided into two great groups: Vertisol and Humic Vertisol, mainly based on the color of the A horizon as indicated in the Vertisolic order chart. The color of the A horizon reflects differences in the nature and amount of organic matter incorporated with the mineral material as a result of differences in climate and vegetation.

Subgroups are separated based on the kind and sequence of horizons indicating conformity with the central concept of the great group. Most of the subgroups within the Vertisolic order were formerly recognized taxonomically at the family or series level as Grumic (Agriculture Canada Expert Committee on Soil Survey 1987).

Vertisol

These are soils that generally occur in the most arid portion of the climatic range of Vertisolic soils. They are primarily associated with a native vegetation of xerophytic and mesophytic grasses and forbs. They have both a vertic and a slickenside horizon as specified for the Vertisolic order. Except for the poorly drained members, they also have brownish-colored A horizons (dry color value of 3.5 or lighter, chroma usually greater than 1.5 dry, and organic C usually 3% or less) similar to associated Chernozemic soils. However, they differ from Chernozemic soils in that it is

Vertisolic Order		
	Vertisol	Humic Vertisol
Color value of A horizon	≥3.5 (dry)	<3.5 (dry)
Chroma of A horizon	usually >1.5 (dry)	usually ≤1.5 (dry)
A horizon	not easily distinguishable from rest of solum	easily distinguishable from rest of solum
	Gleysolic Vertisol	Gleysolic Humic Vertisol
Ah horizon	<10 cm thick	≥10 cm thick

often difficult to separate the A horizon from the rest of the solum based on color alone, and the Ah horizon is massive and hard when dry. They also lack other horizons, in the upper solum, common to Chernozemic or other soils.

The poorly drained members occur commonly in depressional positions associated with better-drained Vertisolic soils. Besides the presence of both a slickenside and a vertic horizon, they have the general characteristics specified for the Gleysolic order. As well, under native conditions they lack a mineral–organic surface (Ah) horizon 10 cm or more in thickness. If cultivated the Ap horizon has one of the following properties:

1. Less than 2% organic C.
2. A rubbed color value greater than 3.5 (moist) or greater than 5.0 (dry).
3. Little contrast in color value with the underlying layer (a difference of less than 1.5 units, if the value of the underlying layer is 4 or more, or a difference of less than 1 unit, if that value is less than 4).

Orthic Vertisol

Common horizon sequence: Ah, Bv or Bvk, Bss or Bssk or Ckss, Ck

These soils have the general properties specified for the Vertisolic order and those members of the Vertisol great group other than the poorly drained members. They have a vertic horizon (Bv or Bvk) and a slickenside horizon (ss), the upper boundary of which occurs within 1 m of the mineral surface. The slickenside horizon may be confined to the B horizon (Bss or Bssk), may form part of the C horizon (Ckss), or may form part of both.

Orthic Vertisols, and all other subgroups of Vertisolic soils, may have saline features. These features are separated taxonomically at the series level, or as a phase of any taxonomic level above the family.

Gleyed Vertisol

Common horizon sequence: Ah, Bvgj or Bvkgj, Bssgj or Bsskgj or Ckssgj, Ckgj or Ckg

These soils have the general properties specified for the Vertisolic order and those members of the Vertisol great group other than the poorly drained members. They differ from Orthic Vertisols by having faint to distinct mottles that indicate gleying within 50 cm of the mineral surface.

Gleysolic Vertisol

Common horizon sequence: LFH or O, Bvg, Bssg or Cgss, Cg

These soils have the general properties specified for the Vertisolic order and the poorly drained members of the Vertisol great group. They differ from Orthic and Gleyed Vertisols mainly by having, within 50 cm of the surface, either colors that indicate poor drainage and periodic reduction (diagnostic of the Gleysolic order) or prominent mottles that indicate gleying, or both.

Humic Vertisol

These Vertisolic soils occur in the more moist portion of the climatic range of Vertisolic soils and are primarily associated with a native vegetation of mesophytic grasses and forbs or with mixed grass, forb, and tree cover. They have an A horizon that is darker in color than soils of the Vertisol great group. Except for the poorly drained members, the surface horizon has a dry color value of less than 3.5, chroma usually 1.5 or less dry, and organic C usually greater than 3%. The A horizon is easily distinguishable from lower horizons based on color alone. However, the boundary between the A horizon and lower horizons is usually very irregular. These soils have both a vertic and a slickenside horizon as specified for the Vertisolic order. They lack other horizons in the upper solum, common to other soils, except they may have a Chernozemic-like A horizon. The poorly drained members occur commonly in depressional positions associated with better-drained Vertisolic soils. Besides the presence of both a slickenside and a vertic horizon, they have the general characteristics specified for the Gleysolic order. As well, under native conditions they have a mineral–organic surface (Ah) horizon 10 cm or more in thickness. If cultivated, the Ap horizon has all the following properties:

1. At least 2% organic C.
2. A rubbed color value of 3.5 or less (moist) or 5.0 or less (dry).
3. At least 1.5 units of color value (moist) lower than that of the next underlying horizon, if the color value (moist) of that horizon is 4 or more, or 1 unit of color value lower than that of the underlying horizon, if its color value is less than 4.

Orthic Humic Vertisol

Common horizon sequence: <u>Ah</u>, <u>Bv</u> or <u>Bvk</u>, <u>Bss</u> or <u>Bssk</u> or <u>Ckss</u>, Ck

These soils have the general properties specified for the Vertisolic order and those members of the Humic Vertisol great group other than the poorly drained members. They have a vertic horizon (Bv or Bvk) and a slickenside horizon (ss), the upper boundary of which occurs within 1 m of the mineral surface. The slickenside horizon may be confined to the B horizon (Bss or Bssk), may form part of the C horizon (Ckss), or may form part of both.

Gleyed Humic Vertisol

Common horizon sequence: <u>Ah</u>, <u>Bvgj</u> or <u>Bvkgj</u>, <u>Bssgj</u> or <u>Bsskgj</u> or <u>Ckssgj</u>, Ckgj or Ckg

These soils have the general properties specified for the Vertisolic order and those members of the Humic Vertisol great group other than the poorly drained members. They differ from Orthic Humic Vertisols by having faint to distinct mottles that indicate gleying within 50 cm of the mineral surface.

Gleysolic Humic Vertisol

Common horizon sequence: LFH or O, <u>Ah</u>, <u>Bvg</u>, <u>Bssg</u> or <u>Cgss</u>, Cg

These soils have the general properties specified for the Vertisolic order and the poorly drained members of the Humic Vertisol great group. They differ from Orthic and Gleyed Humic Vertisols mainly by having either colors that indicate poor drainage and periodic reduction, diagnostic of the Gleysolic order or prominent mottles that indicate gleying within 50 cm of the mineral surface, or both.

Chapter 14

Soil Family and Series

The criteria and guidelines used in differentiating classes in the family and series categories are outlined in this chapter. However, the numerous classes are not defined, particularly for the soil series category (*see* Chapter 17 for definitions of terminology).

The family is a category of the system of soil classification in the same sense as the order, great group, and subgroup. However, the family is not yet as widely used as the long-established categories such as the great group and series. The soil family, as used in *The Canadian System of Soil Classification*, was developed in the 1960s and the first version was adopted in 1968 (National Soil Survey Committee 1968). At that time, terminology and class limits developed in the US Soil Taxonomy were partially adopted but, in some cases, applied somewhat differently to fit the needs of the Canadian system.

Historically, the family category was needed because the number of soil series was too great, and the higher categories too heterogeneous, to be used for many objectives. Therefore, the soil family is used to define and group soil series of the same subgroup, which are relatively uniform in their physical and chemical composition and environmental factors. At the subgroup level all genetic factors are adequately taken care of. At the family level, the practical physical factors that affect plant growth and engineering uses of soils are taken into account. The relative weight of engineering influences versus agronomic influences on the choice of boundaries for family classes is about equal. For example, in the particle-size classes, the limits of 18% clay between coarse and fine and loamy, reflect the change from nonplastic to plastic limit. This is considered by engineers to be an important distinction. Similar breaks occur at the 35% and 60% clay content. On the other hand, there is an important agricultural difference between the coarse and fine silty and loamy classes, especially in terms of capillary rise and available moisture-holding capacity. Basically, the family grouping is intended to allow groupings of soils that have a similar response to management, and to some extent, for engineering and related uses.

Therefore, soils in a family have in common a combination of important specific properties adequate for broad interpretations but inadequate for quantitative interpretations. Soil series are better suited to that purpose. Although the series category has been used throughout the history of soil survey in Canada, it has evolved to an increasingly specific category. Some of the series, which were established before the family category was introduced, can now be divided into several families. In that way, the family level becomes a framework (correlation yardstick) for checking and establishing proper limits for soil series.

Subgroups are divided into families based on certain chemical, physical, and other properties of the soil that reflect environmental factors. The family differentiae are uniform throughout the nine orders of mineral soils. Another set of differentiae is used uniformly for soils of the Organic order. The differentiating criteria for families of mineral soils are particle size, mineralogy, reaction and calcareousness, depth, soil temperature, and soil moisture regime. Those for families of Organic subgroups are characteristics of the surface tier, reaction, soil temperature, soil moisture regime, particle size of terric layer, and the kind and depth of limnic layer. Many of these properties are major ones with respect to the suitability of the soil for various uses. An Orthic Regosol might occur in fragmental or clayey material, or material of some intermediate particle-size class. Particle size, which affects many uses, is not diagnostic of soil classes above the family category. A Rego Black Chernozem soil might have a lithic contact at 15 cm or it might occur in deep unconsolidated material. This important difference is not recognized taxonomically above the family level.

Family Criteria and Guidelines for Mineral Soils

The diagnostic criteria (reaction, calcareousness, and depth classes) apply to the mineral control section as defined in Chapter 2, whereas particle-size and mineralogical classes are defined on a more restrictive control section (*see* Control section of particle-size classes and substitute classes later in this chapter).

Particle size classes

The term "particle size" refers to the grain size distribution of the whole soil including the coarse fraction (>2 mm). It differs from texture, which refers to the fine earth (≤2 mm) fraction only. Also, textural classes are usually assigned to specific horizons, whereas family particle-size classes indicate a composite particle size of a part of the control section that may include several horizons. These particle-size classes may be regarded as a compromise between engineering and pedological classifications. The limit between sand and silt is 74 μm in engineering classifications and either 50 or 20 μm in pedological classifications. The engineering classifications are based on weight percentages of the fraction less than 74 mm, whereas textural classes are based on the ≤2 mm fraction.

The very fine sand fraction, 0.1–0.05 mm, is split in the engineering classifications. The particle-size classes make much the same split but in a different manner. A fine sand or loamy fine sand normally has an appreciable content of very fine sand, but most of the very fine sand fraction is coarser than 74 μm. A silty sediment, such as loess, also has an appreciable amount of very fine sand, but most of it is finer than 74 μm. In particle-size classes the very fine sand is allowed to "float." It is assigned to sand if the texture is fine sand, loamy fine sand, or coarser and to silt if the texture is very fine sand, loamy very fine sand, sandy loam, silt loam, or a finer class.

The particle-size classes defined herein permit a choice of either 7 or 11 classes depending upon the degree of refinement desired. The broad class "clayey", indicating 35% clay or more in the fine earth of defined horizons, may be subdivided into fine-clayey (35–60% clay) and very-fine-clayey (60% or more clay) classes (Figure 41).

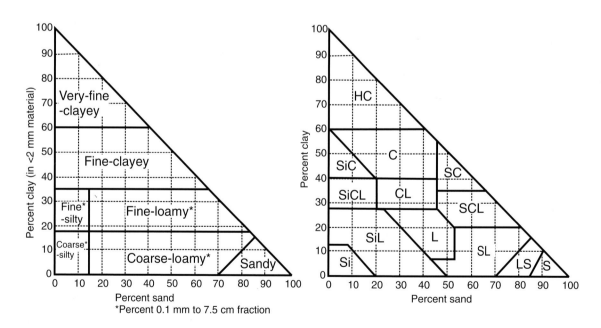

Figure 41 Family particle-size classes triangle (left) and soil texture classes triangle (right). Abbreviations for the texture classes are HC, heavy clay; C, clay; SiC, silty clay; SiCL, silty clay loam; CL, clay loam; SC, sandy clay; SiL, Silt Loam; L, loam; SCL, sandy clay loam; SL, sandy loam; Si, silt; LS, loamy sand; S, sand.

The particle-size classes for family groupings are as follows:

Fragmental Stones (>250mm), cobbles (75–250mm) and gravel (>2–75mm), comprise 90% or more of the soil mass (by volume), with too little fine earth (<10% by volume) to fill interstices larger than 1 mm.

Sandy-skeletal Particles >2 mm occupy 35% or more but less than 90% (by volume), with enough fine earth to fill interstices larger than 1 mm; the fraction ≤2 mm is that defined for the sandy particle-size class.

Loamy-skeletal Particles >2 mm occupy 35% or more but less than 90% (by volume), with enough fine earth to fill interstices larger than 1 mm; the fraction ≤2 mm is that defined for the loamy particle-size class.

Clayey-skeletal Particles >2 mm occupy 35% or more but less than 90% (by volume), with enough fine earth to fill interstices larger than 1 mm; the fraction ≤2 mm is that defined for the clayey particle-size class.

Sandy The texture of the fine earth includes sands and loamy sands, exclusive of loamy very fine sand and very fine sand textures; particles >2 mm occupy less than 35% (by volume).

Loamy The texture of the fine earth includes loamy very fine sand, very fine sand, and finer textures with less than 35% (by weight) clay[1]; particles >2 mm occupy less than 35% (by volume).

- **Coarse-loamy** A loamy particle size that has 15% or more (by weight) of fine sand (0.25–0.1 mm) or coarser particles, including fragments up to 75 mm, and has less than 18% (by weight) clay[1] in the fine earth fraction.

- **Fine-loamy** A loamy particle size that has 15% or more (by weight) of fine sand (0.25–0.1 mm) or coarser particles, including fragments up to 75 mm, and has 18–35% (by weight) clay[1] in the fine earth fraction.

- **Coarse-silty** A loamy particle size that has less than 15% (by weight) of fine sand (0.25–0.1 mm) or coarser particles, including fragments up to 75 mm, and has less

than 18% (by weight) clay[1] in the fine earth fraction.

- **Fine-silty** A loamy particle size that has less than 15% (by weight) of fine sand (0.25–0.1 mm) or coarser particles, including fragments up to 75 mm, and has 18–35% (by weight) clay[1] in the fine earth fraction.

Clayey The fine earth contains 35% or more (by weight) clay[1] and particles >2 mm occupy less than 35% (by volume).

- **Fine-clayey** A clayey particle size that has 35–60% (by weight) clay[1] in the fine earth fraction.

- **Very-fine-clayey** A clayey particle size that has 60% or more (by weight) clay[1] in the fine earth fraction.

Substitute classes for particle-size and mineralogy

Special terms are used for some soils in which particular combinations of texture and mineralogy require special emphasis. At present these include soils containing large amounts of volcanic ash and cinders, and thixotropic[2] soils where particle-size class has little meaning. The terms ashy and cindery apply to some soils differentiated formerly as Andic[3] subgroups of Brunisolic soils. These terms replace both particle-size and mineralogy family class terms.

Cindery At least 60% (by weight) of the whole soil consists of volcanic ash and cinders; 35% or more (by volume) of cinders have a diameter of >2 mm.

Ashy At least 60% (by weight) of the whole soil consists of volcanic ash and cinders; less than 35% (by volume) has a diameter of >2 mm.

Ashy-skeletal Particles >2 mm in diameter, other than cinders, occupy 35% or more (by volume); the fine earth fraction is ashy as defined above.

Thixotropic Particles >2 mm in diameter occupy less than 35% (by volume); the fine earth fraction is thixotropic and the exchange complex is dominated by amorphous materials.

---

[1] Carbonates of clay size are not considered to be clay but are treated as silt.

[2] Property of certain viscous or very humid materials that liquefy after being agitated and resume their initial state when undisturbed.

[3] Soils developed in material rich in glass or where the exchange complex is dominated by amorphous materials.

Thixotropic-skeletal Particles >2 mm in diameter, other than cinders, occupy 35% or more (by volume); the fine earth fraction is thixotropic as defined above.

Application of particle-size classes and substitute classes

In assigning particle-size classes only a segment of the mineral control section as defined in Chapter 2 is commonly used. Surface layers are generally excluded and Bn and Bt horizons are given special emphasis. To apply particle-size classes, use the weighted average particle size of a segment of the control section as defined below. The weighted average can usually be estimated, but in marginal cases the weighted average percentage of one or more size fractions may need to be calculated. This is done by summing the products of size fraction percentage times horizon depth for the applicable segment of the control section and dividing by the total thickness.

If there are strongly contrasting particle sizes, as shown in Table 1, both are used, e.g., fine-loamy over sandy.

The following guidelines indicate the segment of the control section used for establishing soil family particle-size classes:

1. In soils having within 35 cm of the mineral soil surface

    a. a lithic contact, particle size is assessed in all of the mineral material above the lithic contact;

    b. a permafrost layer[1], particle size is assessed in all of the mineral material between the surface and a depth of 35 cm.

2. In other soils lacking a significant Bt or Bn horizon[2], particle size is assessed in that segment of the control section between the lower boundary of an Ap horizon or a depth of 25 cm from the mineral soil surface, whichever is deeper, to either

    a. a depth of 1 m;

    b. a lithic contact; or

    c. a depth of 25 cm below the upper boundary of a permafrost layer, whichever is shallower.

3. In other soils that have a significant Bt or Bn horizon[3] extending deeper than 25 cm from the mineral soil surface the particle size is assessed

    a. in the upper 50 cm of the Bt or Bn horizons (or the entire horizon if thinner), if there are no strongly contrasting particle-size classes in or below these horizons and there is no lithic contact at a depth of less than 50 cm from the top of the Bt or Bn horizon;

    b. in that segment of the control section between the top of the Bt or Bn horizon and the 1 m depth or to a lithic contact, which ever is shallower, if the Bt or Bn horizon contains strongly contrasting particle-size classes;

    c. in the 25–100 cm depth, if there are no strongly contrasting classes in or below the Bt or Bn horizon, but there is a strongly contrasting A horizon more than 50 cm thick.

4. If the base of a significant Bt or Bn horizon, or the segment of the mineral control section in which it occurs, is shallower than 25 cm from the mineral soil surface, particle size is assessed from the lower boundary of that segment below the top of the Bt or Bn or below the base of the Ap horizon, whichever is shallower, to either

    a. a depth of 1 m; or

    b. a lithic contact, whichever is shallower.

Strongly contrasting particle-size classes and substitute classes

These classes identify major variations within the control section that affect properties such as water movement and retention. They emphasize features that may not have been identified at higher taxonomic levels.

The minimum significant thickness of a strongly contrasting layer is 15 cm. The particle-size classes in Table 1 are strongly contrasting if the transition is less than 12 cm thick. For ashy-skeletal and thixotropic-

---

[1] A permafrost layer is a perenially frozen soil layer (MAST <0°C).

[2] For this purpose a significant Bt horizon is at least 15 cm thick and has an upper boundary within a depth of 50 cm of the mineral soil surface.

[3] For this purpose a significant Bt or Bn horizon is at least 15 cm thick and has an upper boundary within a depth of 50 cm of the mineral soil surface.

Table 1. Strongly contrasting particle sizes

| | Fragmental | Sandy-skeletal | Loamy-skeletal | Clayey-skeletal | Sandy | Loamy | | | | Clayey | | Cindery | Ashy | Thixotropic |
						Coarse-loamy	Coarse-silty	Fine-loamy	Fine-silty	Fine-clayey	Very fine-clayey			
						Over[3]								
Fragmental			X[1]		X	X[2]	X	X	X	X	X			X
Sandy-skeletal					X	X	X	X	X	X	X	X		X
Loamy-skeletal	X									X			X	X
Clayey-skeletal	X	X			X									
Sandy	X			X			X	X	X	X	X	X		X
Loamy	X	X		X						X	X	X	X	X
Clayey	X	X	X		X	X	X	X	X					
Cindery													X	

[1] Recognized permutations are designated by an X at the intersect.
[2] The underscore indicates that the broader term (loamy) can be used if desired.
[3] Example usage: Loamy skeletal over Fragmental (X at the intersect).

skeletal classes, enter the table at clayey-skeletal.

Where three strongly contrasting layers occur within the control section, the lowest layer and the thicker of the overlying layers are used to establish contrasting classes.

Strongly contrasting particle-size classes are written as follows: sandy over clayey, fragmental over sandy, etc.

Mineralogy classes

Family mineralogy classes are based on the mineralogical composition of selected particle-size fractions in that segment of the control section used for the designation of the particle-size class. If contrasting particle-size classes are recognized, the mineralogy of only the upper contrasting layer defines the family mineralogy. Like a key, soils are placed in the first of the 13 mineralogy classes defined in Table 2 that accommodates them, even though they may meet the requirements of other classes. Thus a soil that has a $CaCO_3$ equivalent of more than 40% throughout the control section, combined with a mixture of quartz, feldspar, illite, and vermiculite, will be designated as a carbonatic family mineralogy class.

In the absence of data, the placement of soils will commonly depend on judgment. Many of the mineralogy family classes are rare in Canada and relate to specific parent materials. Most Canadian soils have mixed mineralogy; notable exceptions are smectitic, clayey soils of the Interior Plains of western Canada.

Depth classes

Depth classes are applicable only in mineral soils having a lithic contact or permafrost within a depth of 1 m. In the following classes for mineral soils, depth is measured from the mineral soil surface to the contact:

Class	Depth (cm)
Extremely shallow lithic	<20
Very shallow lithic	20–50
Shallow lithic	>50–100
Extremely shallow cryic	<20
Very shallow cryic	20–50
Shallow cryic	>50–100

Reaction classes

It is assumed that the range of pH in the solum is sufficiently well characterized in the subgroup category of most soils and requires no special recognition at the family level. Important differences in reaction in subgroups of Gleysols and Gray Luvisols can be accommodated at the series level. Family reaction classes are applicable only to the C

Table 2 Key to mineralogy classes

Class	Definition	Determinant particle-size fraction
Classes applied to soil families of any particle-size class		
Carbonatic	More than 40% (by weight) carbonates (expressed as $CaCO_3$ equivalent) plus gypsum, and the carbonates are >65% of the sum of carbonates and gypsum	Whole soil, particles ≤2 mm in diameter, or whole soil ≤20 mm, whichever has higher percentages of carbonates plus gypsum
Serpentinitic	More than 40% (by weight) serpentine minerals (antigorite, chrystile, fibrolite, and talc)	Whole soil, particles ≤2 mm in diameter
Gypsic	More than 40% (by weight) of carbonates (expressed as $CaCO_3$ equivalent) plus gypsum, and the gypsum is >35% of the sum of carbonates and gypsum	Whole soil, particles ≤2 mm in diameter, or whole soil ≤20 mm, whichever has higher percentages of carbonates plus gypsum
Sulfurous	Soils containing either iron sulfates, commonly jarosite, if the pH after oxidation is less than 3.5; or more than 0.75% sulfur in the form of poly-sulfides if the soil contains less than three times as much carbonate (expressed as $CaCO_3$ equivalent) as sulfur	Whole soil, particles ≤2 mm in diameter
Classes applied to soil families having a fragmental, sandy, sandy-skeletal, loamy, or loamy-skeletal particle-size class		
Micaceous	More than 40% (by weight)[1] mica	0.02–2 mm
Siliceous	More than 90% (by weight)[1] of silica minerals (quartz, chalcedony, or opal) and other extremely durable minerals that are resistant to weathering	0.02–2 mm
Mixed	All others that have <40% (by weight)[1] of any one mineral other than quartz or feldspars	0.02–2 mm
Classes applied to soil families having a clayey[2] or clayey-skeletal particle-size class		
Kaolinitic	More than 50% (by weight) kaolinite, tabular halloysite, dickite, and nacrite by weight and smaller amounts of other 1:1 or nonexpanding 2:1 layer minerals or gibbsite and <10% (by weight) smectite	≤0.002 mm
Smectitic	More than 50% (by weight) smectite (montmorillonite or nontronite) or a mixture that has more smectite than any other clay mineral	≤0.002 mm
Illitic	More than 50% (by weight) illite (hydrous mica) and commonly >4% K_2O	≤0.002 mm
Vermiculitic	More than 50% (by weight) vermiculite or more vermiculite than any other clay mineral	≤0.002 mm
Chloritic	More than 50% (by weight) chlorite or more chlorite than any other mineral	≤0.002 mm
Mixed	Other soils	≤0.002 mm

[1] Percentages by weight are estimated from grain counts. Usually a count of one or two of the dominant size fractions of a conventional mechanical analysis is sufficient to place the soil.

[2] The clay mineralogy descriptions for clayey soils are based on the weighted average percentage of the particle-size control section of the fine earth fraction (≤2.0 mm).

horizons of mineral soils. They are used in all subgroups except where they would be redundant, as in the Chernozemic and Solonetzic orders, Gray Brown Luvisol, Melanic Brunisol, and Eutric Brunisol great groups, and soils of sulfurous mineralogy family class.

Classes are based on the average pH in 0.01 M $CaCl_2$ (2:1) of the C horizon (C, Ck, Cs, Cg) including IIC, etc., but excluding Csa and Cca. In the absence of a C horizon, the horizon overlying the lithic contact, or 25 cm below the top of a permafrost layer, is used.

Class	pH
Acid	<5.5
Neutral	5.5–7.4
Alkaline	>7.4

Calcareous classes

It is assumed that carbonate (expressed as $CaCO_3$ equivalent) levels in the solum are sufficiently well understood from the subgroup classification of most soils and require no special recognition at the family level. Important differences in carbonate content in subgroups of Gleysols and Gray Luvisols can be accommodated at the series level. Therefore, family calcareous classes are applicable only to C horizons of mineral soils, the mineral horizon overlying a lithic contact, or the mineral material that occurs 25 cm below the top of a permafrost layer, as described under reaction classes. They are used in all soils with Ck or Cca horizons.

Class	$CaCO_3$ equivalent (%)
Weakly calcareous	1–6
Strongly calcareous	>6–40
Extremely calcareous	>40

The class extremely calcareous is redundant in soils with carbonatic mineralogy.

Soil climate classes and subclasses of mineral soils

The soil climate classes and subclasses are applicable to all soils and the criteria used are those of the map *Soil Climates of Canada* (Clayton et al. 1977). In this system soils can be grouped according to soil temperature classes (Table 3) and soil moisture subclasses (Table 4).

Rather than relying upon the map designations for a given area, soil sites need to be individually assessed on the basis of observations of local climatic and micro-climatic variations. Extrapolation from local meteorological station data should allow for any unrepresentative site features such as vegetation and exposure. A useful estimate of mean summer soil temperature (MSST) can be obtained by averaging the three mid-monthly readings of soil temperature at 50 cm taken in July, August, and September.

Family Criteria and Guidelines for Organic Soils

Family criteria apply to the organic control section as defined in Chapter 2.

Characteristics of surface tier

Characteristics of the surface tier may be recognized by using one of the following:

- Organic surface tier; fennic, silvic, sphagnic (each used only for fibric surface tiers), mesic, humic.
- Mineral surface tier[1], 15–40 cm thick; sandy, coarse-loamy, coarse-silty, fine-loamy, fine-silty, clayey.

Reaction classes

Reaction classes are based on the average pH in 0.01 M $CaCl_2$ (4:1) in some part (Euic) or all parts (Dysic) of the organic materials in the organic control section.

Class	pH
Euic	≥4.5
Dysic	<4.5

Soil climate classes and subclasses of organic soils

The soil climate classes and subclasses are applicable to all soils and the criteria used are those of the map *Soil Climates of Canada* (Clayton et al. 1977). In this system soils can be grouped according to soil temperature classes (Table 3) and soil moisture subclasses (Table 4). These classes were designed for well-drained mineral soils in temperate areas.

---

[1] Definitions for the mineral surface tier classes are the same as those of the particle-size classes.

Table 3 Soil temperature classes

Class	Description
Extremely cold	MAST[1] <–7°C Continuous permafrost usually occurs below the active[2] layer within 1 m of the surface Very short growing season, <15 days >5°C Remains frozen within the lower part of the control section Cold to very cool summer, MSST[3] <5°C No warm thermal period >15°C
Very cold	MAST –7–<2°C Discontinuous permafrost may occur below the active layer within 1 m of the surface Soils with Aquic regimes usually remain frozen within part of the control section Short growing season, <120 days >5°C Degree-days >5°C are <555 Moderately cool summer, MSST 5–<8°C No warm thermal period >15°C
Cold	MAST 2–<8°C No permafrost Undisturbed soils are usually frozen in some part of the control section for a part of the dormant season[4] Soils with Aquic regimes may remain frozen for part of the growing season Moderately short to moderately long growing season, 120–220 days >5°C Degree-days >5°C are 555–<1250 Mild summer, MSST 8–<15°C An insignificant or very short, warm thermal period, 0–50 days >15°C Degree-days >15°C are <30
Cool	MAST 5–<8°C Undisturbed soils may or may not be frozen in part of the control section for a short part of the dormant season Moderately short to moderately long growing season, 170–220 days >5°C Degree-days >5°C are 1250–<1720 Mild to moderately warm summer, MSST 15–<18°C Significant very short to short warm thermal period, <120 days >15°C Degree-days >15°C are 30–220
Mild	MAST 8–<15°C Undisturbed soils are rarely frozen during the dormant season Moderately long to nearly continuous growing season, 200–365 days >5°C Degree-days >5°C are 1720–2775 Moderately warm to warm summer, MSST 15–<22°C Short to moderately warm thermal period, <180 days >15°C Degree-days >15°C are 170–670

[1] MAST: mean annual soil temperature.

[2] Segment of the soil seasonally frozen and thawed.

[3] MSST: mean summer soil temperature.

[4] Dormant season <5°C.

Table 4 Soil moisture subclasses

REGIME/Subclass	Description
AQUEOUS	Free water standing continuously on the soil surface
AQUIC	Soil is saturated for significant periods of the growing season
Peraquic	Soil is saturated for very long periods Ground water level is at or within the capillary reach of the surface
Aquic	Soil is saturated for moderately long periods
Subaquic	Soil is saturated for short periods
MOIST UNSATURATED	Varying periods of intensities of water deficits during the growing season
Perhumid	No significant water deficits in the growing season Water deficits <2.5 cm; CMI[1] >84
Humid	Very slight deficits in the growing season Water deficits 2.5-<6.5 cm; CMI[1] 74–84
Subhumid	Significant deficits in the growing season Water deficits 6.5-<13 cm; CMI[1] 59–73
Semiarid	Moderately severe deficits in the growing season Water deficits 13-<19 cm; CMI[1] 46–58
Subarid	Severe deficits in the growing season Water deficits 19-<38 cm in cool and cold regimes; 19-51 cm in mild regimes; CMI[1] 25–45
Arid	Very severe deficits in the growing season Water deficits ≥38 cm in cool regimes and ≥51 cm in mild regimes; CMI[1] <25

[1] The term Climatic Moisture Index (CMI) expresses the growing season (>5°C) precipitation as a percentage of the potential water used by annual crops, when water is readily available from the soil.

$$CMI = \frac{P}{P + SM + IR} \times 100$$

P = growing season precipitation.
SM = water available to crops that is stored in the soil at the beginning of the growing season.
IR = irrigation requirements or water deficit for the growing season.

Therefore, Organic soils in mild regimes may have temperatures equivalent to associated mineral soils. Elsewhere, Organic soils probably are at least one temperature class colder than associated imperfectly to well-drained mineral soils.

The moisture subclasses in Table 4 are defined imprecisely based on of the degree and duration of saturation. Table 5 gives guidelines for selecting the appropriate moisture subclass in organic soils. These criteria apply to the surface tier.

Particle-size classes of terric layer

The particle-size classes that are to be recognized at the family level for mineral material in Terric[1] subgroups of Organic soils are fragmental, sandy, sandy-skeletal, loamy, loamy-skeletal, clayey, and clayey-skeletal.

Limnic layer classes

Limnic layer classes apply only to the Limnic subgroups of Organic soils and are marl, diatomaceous earth, and coprogenous earth. The definitions of these materials may be found in Chapter 2 under "Named layers and materials of Organic soils." Note the exclusion from the Organic order of soils in which mineral sediment, marl, or diatomaceous earth layers thicker than 40 cm occur at the surface or that have mineral sediment, marl, or diatomaceous earth layers thicker than 40 cm within the upper 80 cm of the control section.

Depth classes

Depth classes are applicable only in organic soils having a lithic contact or permafrost within a depth of 160 cm and are measured from the surface to the contact layer.

Class	Depth (cm)
Extremely shallow lithic	10–40
Very shallow lithic	40–100
Shallow lithic	100–160
Extremely shallow cryic	10–40
Very shallow cryic	40–100
Shallow cryic	100–160

Nomenclature for Soil Families

The technical soil family name is descriptive and consists of the subgroup name followed by several adjectives designating the mineral or organic family classes and subclasses, and should be terminated by the term *family*. The classes and subclasses are listed in the following order:

- Mineral soils; particle size, mineralogy, depth, reaction, calcareousness, soil temperature, and soil moisture regime.
- Organic soils; characteristics of surface tier, reaction, soil temperature, soil moisture regime, particle-size of terric layer, limnic material, and depth.

Some of the modifiers are not necessary for some subgroups; for example, the reaction class should not be indicated for Alkaline Solonetzs. Some examples of family names are

- Orthic Humo-Ferric Podzol, coarse-loamy, mixed, acid, cool, perhumid family.
- Orthic Eutric Brunisol, coarse-silty over sandy, mixed, shallow, strongly calcareous, cold humid family.
- Terric Mesisol, humic, dysic, cool, aquic, loamy-skeletal family.
- Limnic Humisol, humic, euic, mild, aquic, coprogenous family.

A family thus described is a taxonomic entity within which from one to a large number of series may be established. Like the series, its suitability as a basis for naming pure or complex mapping units varies from region to region and according to the scale of mapping.

In some instances it is useful to indicate phases of families (*see* Chapter 15, Soil Phase). This is done by adding, after the term family, the appropriate phase terms and "phase." An example is

- Orthic Humo-Ferric Podzol, coarse-loamy, mixed, acid, cool, perhumid family; peaty, level phase.

For convenience and brevity the name of a common series may be used to designate a family. For example, it is acceptable to refer to "Breton family" to indicate the Orthic Gray Luvisol, fine-loamy, mixed, neutral, cold, subhumid family.

---

[1] An unconsolidated mineral layer at least 30 cm thick beneath the surface tier.

Table 5 Moisture subclasses as applied to organic soils

Soil moisture regime	Aquic				Moist soils	
Classification	Aqueous	Peraquic	Aquic	Subaquic	Perhumid	Humid
Descriptive condition	Free surface water	Saturated for very long periods	Saturated for moderately long periods	Saturated for short periods	Moist with no significant seasonal deficit	Moist with condition no significant seasonal deficit
Drainage class		Very poorly drained	Poorly drained	Imperfectly drained	Imperfectly to moderately well drained	Moderately well drained
Suggested criteria						
Saturated period	Continuous	Very long	Long to moderately short	Short to very short	Very short	Very short to insignificant
Months per year	11.5–12	>10	4–10	<4	<2	<0.5
Moist period	insignificant	Very short	Short to moderately long	Long to very long	Long to very long	Very long
Months per year	<0.5	<2	2–8	8–11.5	8–11.5	>11.5
Associated native vegetation	Hydrophytic	Hydrophytic	Hydrophytic to mesophytic	Hydrophytic to mesophytic	Mesophytic	Mesophytic
	Nymphaea, Potamogeton, Scirpus, Typha, Phragmites, Drepanocladus	*Scirpus, Typha, Carex, Drepanocladus,* Feather mosses, *Larix*	Wet black spruce forest, mixed feather and sphagnum mosses, Ericaceous shrubs	Wet to very moist black spruce forest, sphagnum mosses, Ericaceous shrubs	Moist black spruce forest, mixed sphagnum and feather mosses, Ericaceous shrubs, lichens	Disturbed species Cultivated species
Associated peat landform	Wetlands, marsh, floating fen, collapse scars	Flat fens, patterned fens, spring fens, swamps	Blanket bogs, transitional bogs	Domes bogs, plateaus	Frozen plateaus, frozen palsas, frozen peat polygons	Drained peat land, Folisols

Soil Series

The concept of the soil series has changed greatly since the early 1900s when a series was somewhat analogous to a geological formation. Now the series is a category in the system of soil taxonomy in the same way that order, great group, subgroup, and family are categories. A soil series is a conceptual class that has, or should have, defined limits in the same way as a great group. The link between the conceptual entity, soil series, and real bodies of soil is the pedon. Any pedon may be classified as a unique soil series, but series have been named for only a very small proportion of the kinds of pedons that occur.

Soil series are subdivisions of soil families based upon relatively detailed properties of the pedon within the depth of the control section. The range of variability of the differentiating characteristics is narrower for the series than for the family. Series cannot transgress soil climatic and particle-size classes, or other boundaries recognized in family separations. The significance of differences in the properties of the different kinds of pedons that fall within a soil family depends on how these properties combine. No specific property, or group of properties, has been assigned limits and been used consistently from family to family and within families to define series. Each potential soil series is treated as an individual case and the decision on whether or not it should be recognized as a separate taxon involves a judgment based on the following guidelines:

1. The properties that distinguish a particular series from other series must be sufficiently recognizable that qualified pedologists can identify the series consistently.

2. The properties used to differentiate series must be within the control section (*see* 4 and 5 below).

3. Soils of a series must occupy at least a few hundred hectares. Establishing a series to classify a few pedons that occupy a few hectares is not justified even if the pedons have unique properties.

4. Soil series within families of mineral soils are usually differentiated based on the following properties:

    a. color, including mottling;

    b. texture;

    c. structure;

    d. consistence;

    e. thickness, relative arrangement of horizons, and degree of expression of horizons and of the solum;

    f. abundance and size of coarse fragments;

    g. depth to a lithic contact, permafrost, or contrasting material to a finer degree than used in higher categories;

    h. depth to, and concentration of free carbonates;

    i. depth to, and concentration of soluble salts;

    j. reaction (pH);

    k. lithology.

5. Soil series within families of Organic soils may be differentiated based on the following properties:

    a. material composition— botanical origin of fibers and nature of terric layer, if any;

    b. thickness, amount of decomposition and relative arrangement of layers;

    c. abundance of woody material—logs and stumps;

    d. calcareousness;

    e. bulk density;

    f. mineral content of organic material;

    g. soil development in the terric layer;

    h. mineralogy of terric or cumulic layers;

    i. texture of terric or cumulic layers;

    j. reaction (pH).

Few series of Organic soils have been established and it is likely that other series criteria will emerge.

Pedons classified as a given soil series have a similar number and arrangement of horizons whose color, texture, structure, consistence, thickness, reaction, or some combination of these properties are within a defined range. In the case of soils without genetic horizons, the above statement applies to the C horizons to the depth of the control section.

The concept of the soil series has been refined progressively in Canada throughout the last half century. Many "series" established 30 or more years ago might include pedons that belong to several subgroups or families today. Years ago soil taxonomy was focused on the series and the great group; much less attention was given to other categories. Series were differentiated without reference to family criteria, which were not developed until recently. Thus many of the "series" used today still include, to a degree, the attributes of the more generalized series of several years ago. In the process of establishing new series and refining old series today, the pedologist should work downward in soil taxonomy considering the differentiation of soil properties at the order, great group, subgroup, and family levels before subdividing the family into series. Taxonomy will probably not be extended to the series level in many medium- to small-scale soil surveys. For more detailed work, the series is a category of paramount importance because it is the most specific level in soil taxonomy and the one used for most interpretations. Sound judgments, based upon the guidelines stated, on the part of soil mappers and correlators are essential in decisions on establishing series. The definition of a series implies a statement of the limits of its properties.

Chapter 15

Soil Phase

A soil phase is a unit of soil outside the system of soil taxonomy. It is a functional unit that may be designed according to the purpose of the survey. Phases of taxa at any categorical level, from order to series, may be defined. Also, areas not classified in soil taxonomy such as rockland and steep slopes may be designated as phases on soil maps. The two general reasons for differentiating soil phases are

- to recognize and name soil and landscape properties that are not used as criteria in soil taxonomy, for example, slope or erosion

- to recognize and name, at a relatively high categorical level, soil properties that are used as differentiae at a lower categorical level. For example, depth to a lithic layer is a family criterion, but it can be used as a phase criterion at the order, great group, and subgroup levels such as, Brunisolic soils, very shallow lithic phase.

The properties recognized above must be associated with areas of soil or nonsoil as mapped. The major phase differentiae are listed below.

Slope

The slope classes are defined as follows:

Slope class	Percent slope	Approximate degrees	Terminology
1	0–0.5	0	level
2	>0.5–2	0.3–1.1	nearly level
3	>2–5	>1.1–3	very gentle slopes
4	>5–10	>3–5	gentle slopes
5	>10–15	>5–8.5	moderate slopes
6	>15–30	>8.5–16.5	strong slopes
7	>30–45	>16.5–24	very strong slopes
8	>45–70	>24–35	extreme slopes
9	>70–100	>35–45	steep slopes
10	>100	>45	very steep slopes

For example, Dystric Brunisol and rock outcrop, moderate slopes.

Water Erosion

The following water-erosion classes as defined in the *Soil Survey Manual* of the U.S. Department of Agriculture (Soil Survey Staff 1951, pp. 261–264) are used as phases.

Class W1, slightly eroded phase

As much as 25% of the original A horizon may have been removed from most of the area. In most cases the soils eroded to this degree are not significantly different in use capabilities and management requirements from noneroded soils.

Class W2, moderately eroded phase

Between 25 and 75% of the original A horizon may have been lost from most of the area. The present Ap horizon consists of a mixture of the underlying soil and the original A horizon. Shallow gullies may be present.

Class W3, severely eroded phase

More than 75% of the original A horizon and commonly part of the next underlying horizon have been lost from most of the area. Shallow gullies are common and a few deep ones may occur.

Class W4, gullied land phase

The land is dissected by moderately deep to deep gullies with small areas of intact soil between the gullies. The area is unsuitable for crop production without reclamation.

Wind Erosion

The following wind erosion classes, as defined in the *Soil Survey Manual* of the U.S. Department of Agriculture (Soil Survey Staff 1951, p. 267), are used as soil phases.

Class D1, slightly wind-eroded phase

Wind has removed between 25 and 75% of the original A horizon and tillage results in mixing of subsurface material with remnants of the original surface layer.

Class D2, severely wind-eroded phase

Wind has removed more than 75% of the original A horizon and commonly part of the underlying horizon.

Class D3, blown-out land phase

Wind has removed most of the solum and numerous blowout holes are carved into the parent material. Some areas between blowouts are deeply buried by soil material from the blowouts. The area is unsuitable for crop production without extensive reclamation.

Soil Deposition

The phases for deposition as described in the *Soil Survey Manual* of the U.S. Department of Agriculture (Soil Survey Staff 1951, pp. 295–296), are being used currently in Canada. Two phases are defined as follows:

Overblown phase

Deposits of wind-eroded materials on the soil surface are great enough to influence management but are not great enough to destroy the essential characteristics of the soil series.

Overwash phase

Deposits of water-eroded materials on the soil surface are thick enough to influence management requirements significantly but are not deep enough to destroy the essential characteristics of the soil series.

Stoniness

The phases for stoniness are described in the *Soil Survey Manual* of the U.S. Department of Agriculture (Soil Survey Staff 1951, pp. 216–220). Six phases of stoniness are defined on the basis of the percentage of the land surface occupied by fragments coarser than 25 cm in diameter.

Class S0, nonstony phase

No stones or too few are present to interfere with cultivation (<0.01% of surface, stones more than 25 m apart).

Class S1, slightly stony phase

Some stones are present that hinder cultivation slightly or not at all (0.01–0.1% of surface, stones 8–25 m apart).

Class S2, moderately stony phase

Enough stones are present to cause some interference with cultivation (0.1–3% of surface, stones 1–8 m apart).

Class S3, very stony phase

There are sufficient stones to handicap cultivation seriously; some clearing is required (3–15% of surface, stones 0.5–1 m apart).

Class S4, exceedingly stony phase

The stones prevent cultivation until considerable clearing is done (15–50% of surface, stones 0.1–0.5 m apart).

Class S5, excessively stony phase

The land surface is too stony to permit cultivation; it is boulder or stone pavement (more than 50% of surface, stones less than 0.1 m apart).

Rock Outcrop

Six phases of rockiness (bedrock exposure) are defined as follows:

Class R0, nonrocky phase

Bedrock exposures do not interfere seriously with tillage. Exposures, if present, are generally more than 100 m apart and cover less than 2% of the surface.

Class R1, slightly rocky phase

The bedrock exposures interfere with tillage but not enough to make intertilled crops impracticable. Depending on the pattern and how it affects tillage, rock exposures are roughly 35–100 m apart and cover 2–10% of the surface.

Class R2, moderately rocky phase

The bedrock exposures make tillage of intertilled crops impracticable, but the soil can be worked for hay crops or improved pasture if other soil characteristics are favorable. Rock exposures are roughly 10–35 m apart and cover about 10–25% of the surface depending on the pattern.

Class R3, very rocky phase

The rock outcrops make all use of machinery impracticable, except for small machinery. Where other soil characteristics are favorable the land may have some use for native pasture or forests. Rock exposures or patches of soil too thin over rock for use are roughly 3.5–10 m apart and cover 25–50% of the surface depending on the pattern.

Class R4, exceedingly rocky phase

Sufficient rock outcrop or insufficient depth of soil over rock makes all use of machinery impracticable. The land may have some value for poor pasture or forestry. Rock outcrops are less than 3.5 m apart and cover 50–90% of the area.

Class R5, excessively rocky phase

More than 90% of the land surface is exposed bedrock (rock outcrop).

Folic

Any mineral soil having a surface horizon of 15–40 cm of folic material may be designated as a folic phase.

Peaty

Any mineral soil having a surface horizon of 15–60 cm of fibric organic material or 15–40 cm of mesic or humic organic material may be designated as a peaty phase.

Cryic

Any noncryoturbated mineral or organic soil having permafrost below the 1 m depth, or cryoturbated mineral soil having permafrost below the 2 m depth, may be designated as a cryic phase.

Cryoturbated

Any nonpermafrost soil having one or more cryoturbated horizons may be designated as a cryoturbated phase.

Other Differentiae

Other differentiae that are used taxonomically at lower categorical levels are principally family and some series criteria. These may be used as phase criteria at the order, great group, and subgroup levels.

Family criteria

Definitions are given in Chapter 14.

Particle size e.g., Humo-Ferric Podzols, fragmental phase. The textural class of the mineral surface layer of a soil series may also be indicated as a phase, e.g., Breton silt loam.

Substitute classes (particle size) e.g., Dystric Brunisols, cindery phase.

Mineralogy e.g., Black, smectitic phase.

Depth e.g., Regosols, shallow lithic phase (lithic contact 50–100 cm from the mineral surface); Organic Cryosols, very shallow cryic phase (permafrost layer at 40–100 cm).

Reaction e.g., Regosols, alkaline phase.

Calcareous e.g., Rego Dark Brown Chernozem, extremely calcareous phase.

Soil climate e.g., Podzolic, cold, perhumid phase.

Other family criteria for Organic soils that may be used as phases of higher categories are characteristics of the surface tier, e.g., Mesisols, fine loamy phase; reaction, e.g., Typic Fibrisols, dysic phase; and kind of limnic material, e.g., Limnic Mesisol, diatomaceous phase.

Series criteria

Physical disruption e.g., Humo-Ferric Podzol, turbic phase.

Salinity e.g., Orthic Brown Chernozem, saline phase.

Influence of volcanic ash   e.g., Orthic Dystric Brunisol, andic phase.

Secondary carbonates in the A horizon   e.g., Orthic Brown Chernozem, carbonated phase.

Other series differentiae may also be named as phases of higher categories.

Also, subgroup differentiae may be used to indicate phases of classes at the order level, e.g., Podzolic soils, gleyed phase.

Chapter 16

Correlation of Canadian Soil Taxonomy with Other Systems

The approximate equivalents of the soil horizons and taxa in the Canadian system are given for the U.S. system (Soil Survey Staff 1994) and in the terminology of the FAO-UNESCO soil map of the world (Food and Agriculture Organization of the United Nations 1985). The horizon designations and terms are rarely exactly equivalent. The definitions of soil horizons and soil taxa differ from one system to another.

The U.S. system includes a category named suborder that is not in the Canadian system, and the FAO soil units are arranged into only two categories. Thus the categorical levels of the related taxa are generally not equivalent. The closest approximation to equivalence of taxa is in the case of Organic soils (Canadian) and Histosols (U.S.). Great groups and subgroups of Organic soils are nearly equivalent to the corresponding suborder and great groups of Histosols. This correspondence resulted from United States and Canadian pedologists working together to develop taxonomy for Organic soils. However, even in this case there are differences. For example, the organic matter content required for Histosols differs somewhat from that required for soils of the Organic order.

In other orders, differences in the approximately equivalent taxa are greater. For example, most Chernozemic soils are Mollisols, but some are Aridisols. Many Mollisols are Chernozemic soils, but some are Gleysolic soils, others are Solonetzs, and still others are Melanic Brunisols. There are also basic differences between Podzolic soils and Spodosols. Most Podzolic soils are Spodosols, but a significant proportion are Inceptisols because a spodic horizon must be dominated more by amorphous material than a podzolic B horizon. Conversely, a few Podzolic Luvisols are Spodosols. All Vertisolic soils would qualify as Vertisols. However, all Vertisols would not be Vertisolic because Vertisolic soils must have a layer affected by argillipedoturbation as well as slickensides and cracks. At lower levels in the systems, differences become progressively greater.

Tables 6, 7, and 8 provide some idea of how soil horizon designations and soil taxa are related at the upper levels in the three systems. The tables have been revised from those prepared by J.S. Clayton for previous editions of the *Canadian System of Soil Classification*. Revisions were required because of some changes in horizon designations and in the definitions of horizons and taxa in both the Canadian and U.S. systems. Correlation at the subgroup level is not presented.

The tables are not adequate for correlating either the horizon designations or the taxonomy of a given pedon as expressed in one system with those of another system. Adequate correlation requires knowing the criteria and definitions in the systems involved.

Table 6 Correlation of horizon definitions and designations

1. Canadian	2. U.S.	3. FAO	Comments
O	O	⎡H	1 (Can. limit): organic horizon (O) >17% organic C*
Of	Oi	⎢H	2 & 3 (U.S. and FAO limits): lower limit of organic
Om	Oe	⎢H	horizons ranges proportionately from 20% OM with
Oh	Oa	⎣H	0% clay to 30% OM with >50% clay
Oco	Oa	H	Coprogenous limnic material
L-F	Oi-Oe	O	Generally not saturated with water for prolonged periods
L-H	Oi-Oa	O	
F-H	Oe-Oa	O	
A	A	⎡A	1: ≤17% organic C; 2 & 3: upper limit of OM ranges
Ah	A	⎢Ah	proportionately from 20% OM with 0% clay to 30% OM
Ahe	AE	⎣(Ah-E)	with clay >50%
Ae	E	E	
Ap	Ap	Ap	
AB	AB or EB	AB or EB	Transitional horizons
BA	BA or BE	BA or BE	
A & B	A & B	A/B	Interfingered horizons
AC	AC	A/C	
B	B	B	
Bt	Bt	Bt	
Bf	Bs	Bs	1: specific limits; 2 & 3: no specific limit
Bhf	Bhs	Bhs	1: >5% organic C
Bgf	Bgs	Bgs	
Bh	Bh	Bh	1: specific C to Fep ratio; 2 & 3: no specific C to Fep ratio
Bn	Bn	Bn	
Bm	Bw	Bw	
C	C	C	
IIC	2C	IIC	
R	R	R	
W	—	—	Water
Other suffixes			May be used with A, B, or C horizons
b	b	b	1 & 2: buried; 3: buried or bisequa
c	m	m	
ca	k	k	Accumulation of carbonates
—	y	y	Accumulation of gypsum
cc	m	c	
g	g	g or r	3: g — mottling, — strong reduction
j	—	—	
k	—	—	Indicates presence of carbonate
—	v	—	Plinthite
—	q	q	Silica accumulation
s	z	z	Visible salts
ss	ss	—	Indicates presence of slickensides
sa	y or z	y or z	1: includes gypsum
			2 & 3: y — gypsum, z — other more soluble salts
—	o	—	Residual sequioxide concentration
u	—	—	Turbic
—	—	u	Unspecified
v	—	—	Vertic horizon
x	x	x	Fragipan
y	—	—	Cryoturbation
z	f	i	Permafrost layer

* 17% organic C equals about 30% organic matter.

Table 7 Correlation of United States and FAO diagnostic horizons with nearest Canadian equivalents

1. U.S.	2. FAO	3. Canadian	Comments
Mollic Epipedon	Mollic A	Chernozemic A	With high base status
Anthropic Epipedon	Mollic A	Cultivated Chernozemic A	
Umbric Epipedon	Umbric A	Ah	With low base status
Histic Epipedon	Histic H	Of, Om, Oh	
Ochric Epipedon	Ochric A	light-colored A	
Plaggen Epipedon	—	Ap	
Albic horizon	Albic E	Ae	
Argillic horizon	Argillic B	Bt	
Agric horizon	Argillic B	Illuvial B	Formed under cultivation
Natric horizon	Natric B	Bn or Bnt	
Spodic horizon	Spodic B	podzolic B	
Cambic horizon	Cambic B	Bm, Bg, Btj	
Oxic horizon	Oxic B	—	
Duripan	m	c	
Durinodes	—	cc	
Fragipan	x	Fragipan	
Calcic horizon	Calcic horizon	Bca or Cca	
Petrocalcic	Bkm	Bcac or Ccac	
Gypsic	Gypsic	Asa, Bsa, Csa	3: only if sa horizon is dominantly $CaSO_4$
Salic	—	Asa, Bsa, Csa	
Placic	Thin iron pan	Placic	
Plinthite	Plinthite	—	
Lithic contact	—	Lithic contact	
Paralithic contact	—	IICc	
g	Gleyic horizon	g	
—	sulfuric horizon	—	Low pH, jarosite mottles

Table 8 Taxonomic correlation at the Canadian order and great group levels[1]

Canadian System	U.S. SOIL TAXONOMY	WRB(FAO)[2] System
Chernozemic	Borolls	Kastanozem, Chernozem, Greyzem, Phaeozem
Brown Chernozem	Aridic Boroll subgroups	Kastanozem (aridic)
Dark Brown Chernozem	Typic Boroll subgroups	Kastanozem (Haplic)
Black Chernozem	Udic Boroll subgroups	Chernozem
Dark Gray Chernozem	Boralfic Boroll subgroups, Albolls	Greyzem
Solonetzic	Natric great groups, Mollisols & Alfisols	Solonetz
Solonetz	Natric great groups, Mollisols & Alfisols	Mollic, Haplic, or Gleyic Solonetz
Solodized Solonetz	Natric great groups, Mollisols & Alfisols	Mollic, Haplic, or Gleyic Solonetz
Solod	Glossic Natriborolls, Natralbolls	Solodic Planosol
Vertic Solonetz	Haplocryerts	Sodic Vertisol
Luvisolic	Boralfs & Udalfs	Luvisol
Gray Brown Luvisol	Hapludalfs or Glossudalfs	Albic Luvisol, Haplic Luvisol
Gray Luvisol	Boralfs	Albic Luvisol, Gleyic Luvisol
Podzolic	Spodosols, some Inceptisols	Podzol
Humic Podzol	Cryaquods, Humods	Humic Podzol
Ferro-Humic Podzol	Humic Cryorthods, Humic Haplorthods	Orthic Podzol

(continued)

Table 8 Taxonomic correlation at the Canadian order and great group levels[1] (*concluded*)

Canadian System	U.S. SOIL TAXONOMY	WRB(FAO)[2] System
Humo-Ferric Podzol	Cryorthods, Haplorthods	Orthic Podzol
Brunisolic	Inceptisols, some Psamments	Cambisol
Melanic Brunisol	Cryochrepts, Eutrochrepts, Hapludolls	Cambisol, Eutric Cambisol
Eutric Brunisol	Cryochrepts, Eutrochrepts	Eutric Cambisol, Calcic Cambisol
Sombric Brunisol	Humbric Dystrochrepts	Dystric Cambisol, Umbric Cambisol
Dystric Brunisol	Dystrochrepts, Cryochrepts	Dystric Cambisol
Regosolic	Entisols	Fluvisol, Regosol
Regosol	Entisols	Regosol
Humic Regosol	Entisols	Fluvisol, Regosol
Gleysolic	Aqu-suborders	Gleysol, Planosol
Humic Gleysol	Aquolls, Humaquepts	Mollic, Umbric, Calcic Gleysol
Gleysol	Aquents, Fluvents, Aquepts	Eutric, Dystric Gleysol
Luvic Gleysol	Argialbolls, Argiaquolls, Aqualfs	Planosol
Organic	Histosols	Histosol
Fibrisol	Fibrists	Histosol
Mesisol	Hemists	Histosol
Humisol	Saprists	Histosol
Folisol	Folists	Histosol
Cryosolic	Gelisols	Cryosol
Turbic Cryosol	Turbels	Cryosol
Static Cryosol	Orthels	Cryosol
Organic Cryosol	Histels	Cryic Histosol
Vertisolic	Cryerts	Vertisol
Vertisol	Haplocryerts	Calcic Vertisol, Eutric Vertisol
Humic Vertisol	Humicryerts	Dystric Vertisol

[1] Only the nearest equivalents are indicated.

[2] *World Reference Base for Soil Resources* (Deckers et al. 1998)

Chapter 17

Terminology for Describing Soils

This chapter provides a brief summary of the main terminology used to describe soils at the scales of the landscape and pedon. It refers to the section of the *Canadian Soil Information System (CanSIS) Manual for Describing Soils in the Field 1982 Revised* (Agriculture Canada Expert Committee on Soil Survey 1983) in which the terminology and methods of coding descriptive data are defined in detail.

Landform and relief *See* section 8B in the Manual and Chapter 18 in this publication.

Erosion *See* section 8H in the Manual and Chapter 15 in this publication for definitions of water and wind erosion classes.

Stoniness *See* section 8J in the Manual and Chapter 15 herein for definitions of stoniness classes. Terminology for coarse fragments is given in Table 9.

Rockiness (bedrock exposure) *See* section 8K in the Manual and Chapter 15 herein for definitions of rockiness classes.

Soil water regime *See* sections 8D, D1, D2, E, F, and G in the Manual. The following aspects of the soil water regime are classified: Soil Drainage, Aridity, Hydraulic Conductivity, Impeding Layer, Depth of Saturated Zone and Duration, and Man-made Modifiers.

Soil color *See* section 10C in the Manual. Munsell notations, e.g., 10YR 5/3, (hue, value, and chroma) as well as the appropriate color name (shown for the notation given) are used to indicate the colors of individual horizons of the pedon. Preferably record both moist (10YR 3/3 m) and dry (10YR 5/3 d) soil colors and indicate whether the sample is moist (m) or dry (d) if color is recorded at only one moisture state.

Soil texture *See* section 10K in the Manual. Textural classes are defined in terms of the size distribution of primary particles as estimated by sieve and sedimentation analysis. The textural classes are indicated in Figure 42; named size classes of primary particles and their dimensions are as follows:

Name of separate	Diameter (mm)
very coarse sand	2.0–1.0
coarse sand	1.0–0.5
medium sand	0.5–0.25
fine sand	0.25–0.10
very fine sand	0.10–0.05
silt	0.05–0.002
clay	≤0.002
fine clay	≤0.0002

Table 9 Terminology for various shapes and sizes of coarse fragments

Shape and kind of fragments	Size and name of fragments		
	Up to 8 cm in diameter	*8–25 cm in diameter*	*>25 cm in diameter*
Rounded and subrounded fragments (all kinds of rocks)	Gravelly	Cobbly	Stoney (or bouldery)[1]
Irregularly shaped angular fragments			
Chert	Cherty	Coarse cherty	Stony
Other than chert	Angular gravelly	Angular cobbly	Stony
	Up to 15 cm in length	*15–38 cm in length*	*38 cm in length*
Thin flat fragments			
Thin flat sandstone, limestone, and schist	Channery	Flaggy	Stony
Slate	Slaty	Flaggy	Stony
Shale	Shaly	Flaggy	Stony

[1] Bouldery is sometimes used where stones are larger than 60 cm.

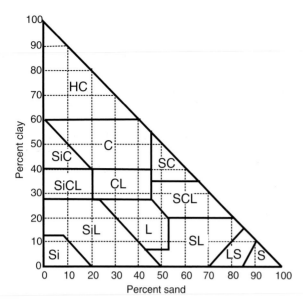

Figure 42 Soil texture classes triangle. Percentages of clay and sand in the main textural classes of soil; the remainder of each class is silt. Abbreviations for the texture classes are: HC, heavy clay; C, clay; SiC, silty clay; SiCL, silty clay loam; CL, clay loam; SC, sandy clay; SiL, Silt Loam; L, loam; SCL, sandy clay loam; SL, sandy loam; Si, silt; LS, loamy sand; S, sand.

Mottles *See* section 10L in the Manual. Mottles are spots or blotches of different color, or shade of color, interspersed with the dominant soil color. Note the color of the matrix and the principal mottles, and the pattern of mottling. The latter is indicated in terms of abundance (few, common, many), size (fine, medium, coarse), and contrast with the matrix (faint, distinct, prominent).

Soil structure *See* section 10M in the Manual. Soil structure refers to the aggregation of primary soil particles into compound particles that are separated from adjoining aggregates by surfaces of weakness. Soil structure is classified in terms of grade or distinctness (weak, moderate, strong), class or size (fine, medium, coarse, very coarse), and type (granular, platy, prismatic, blocky). *See* Table 10 and Figure 43.

Consistence *See* section 10N in the Manual. Soil consistence refers to the soils resistance to deformation or rupture and its degree of cohesion and adhesion. Consistence of wet soil is classified in terms of stickiness (nonsticky, slightly sticky, sticky, or very sticky) and plasticity (nonplastic, slightly plastic, plastic, or very plastic). Consistence is classified for moist soil as loose, very friable, friable, firm, or very firm. For dry soil consistence is classified as loose, soft, slightly hard, hard, very hard, extremely hard, or rigid. Cementation refers to brittle, hard consistence due to some cementing substance. The classes of cementation are weakly cemented, strongly cemented, and indurated.

Roots *See* section 10T in the Manual. Abundance, size, orientation, distribution, and depth of root penetration are noted.

Pores *See* section 10U in the Manual. Abundance, size, orientation, distribution, continuity, morphology, and type of pore are estimated and noted.

Clay films (argillans) *See* section 10S in the Manual. Clay films are described in terms of frequency, thickness, location, and color.

Horizon boundary *See* section 10P in the Manual. The lower boundary of each horizon is described in terms of distinctness (abrupt, clear, gradual, or diffuse) and form (smooth, wavy, irregular, or broken).

Ice *See* Pihlainen and Johnston (1963), Brown and Kupsch (1974), and Harris et al. (1988) for terminology that describes ice and other features of permafrost soils.

Other features *See* sections 8 and 10 in the Manual. Some other features of soils and sites that may be described are humus form, permafrost, land use, concretions, nodules, calcareousness, salinity, coarse fragments, and reaction.

Example of a Pedon Description

Descriptive data for soils are commonly entered on CanSIS forms, but conventional descriptions are needed for some purposes. The following order of listing properties is recommended: color, texture, mottles, structure, consistence, roots, pores, clay films, concretions, carbonates, salts, coarse fragments, horizon boundary, thickness range, and reaction. A description of a Gleyed Humo-Ferric Podzol follows:

Horizon	Depth cm	Description
L-H	7.5–0	Black (10YR 2/1 m), dark grayish brown (10YR 4/1 d); semi-decomposed organic matter; fibrous, abundant, fine and medium roots; abrupt, smooth boundary; 5–10 cm thick; acid.
Ae	0–10	Gray (5YR 6/1 m), light gray (5YR 7/1 d); sandy loam; single grain; loose, friable; few, fine and medium roots; few, fine, vesicular pores; clear, wavy boundary with some fine tongues into underlying horizon; 5–12 cm thick; acid.
Bfgj	10–30	Reddish brown (5YR 4/4 m, 5/4 d); sandy loam; common, medium, distinct strong brown (7.5 YR 5/6) mottles; amorphous; friable; few, fine and very fine roots; few, medium and fine pores; some gravel; clear, smooth boundary; 15–25 cm thick; acid.
Bfg	30–61	Reddish brown (5YR 4/3 m, 5/3 d); sandy loam; many, medium to coarse, prominent strong brown (7.5 YR 5/6) mottles; amorphous; firm; few stones; clear, smooth boundary; 20–38 cm thick; acid.
C	61+	Reddish brown (2.5YR 4/4 m, 5/4 d) sandy loam; amorphous; firm; slightly plastic; some stones; acid.

Table 10 Types and classes of soil structure

Type	Kind	Class	Size (mm)
Structureless: no observable aggregation or no definite orderly arrangement around natural lines of weakness	**Single grain structure**: loose, incoherent mass of individual particles as in sands **Amorphous (massive) structure**: a coherent mass showing no evidence of any distinct arrangement of soil particles		
Blocklike: soil paticles are arranged around a point and bounded by flat or rounded surfaces	**Blocky (angular blocky)**: faces rectangular and flattened, vertices sharply angular	Fine blocky Medium blocky Coarse blocky Very coarse blocky	<10 10–20 20–50 >50
	Subangular blocky: faces subrectangular, vertices mostly oblique, or subrounded	Fine subangular blocky Medium subangular blocky Coarse subangular blocky Very coarse subangular blocky	<10 10–20 20–50 >50
	Granular: spheroidal and characterized by rounded vertices	Fine granular Medium granular Coarse granular	<2 2–5 5–10
Platelike: soil particles are arranged around a horizontal plane and generally bounded by relatively flat horizontal surfaces	**Platy structure**: horizontal planes more or less developed	Fine platy Medium platy Coarse platy	<2 2–5 >5
Prismlike: soil particles are arranged around a vertical axis and bounded by relatively flat vertical surfaces	**Prismatic structure**: vertical faces well defined, and edges sharp	Fine prismatic Medium prismatic Coarse prismatic Very coarse prismatic	<20 20–50 50–100 >100
	Columnar structure: vertical edges near top of columns not sharp; columns flat-topped, round-topped or irregular	Fine columnar Medium columnar Coarse columnar Very coarse columnar	<20 20–50 50–100 >100

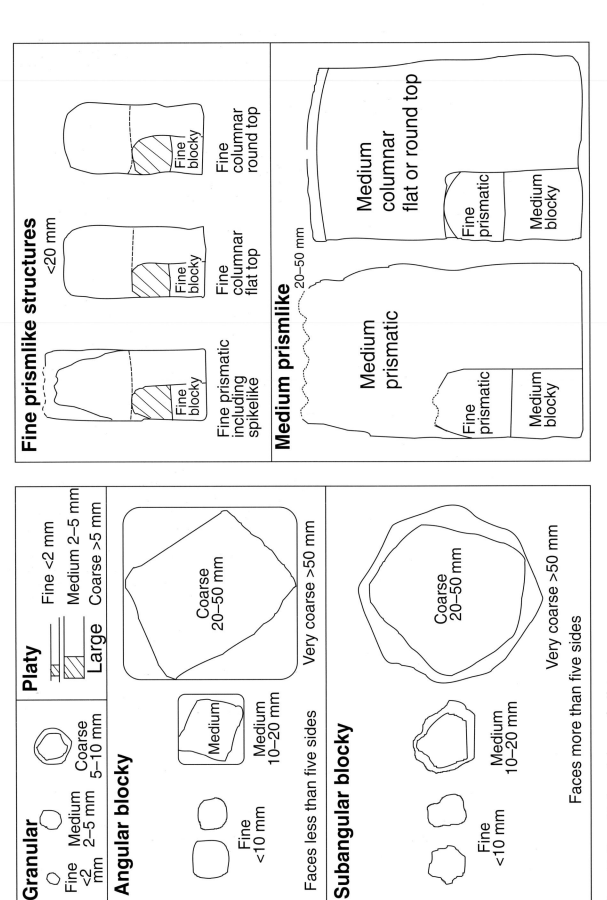

Fine prismlike structures
<20 mm

Fine prismatic
including
spikelike

Fine
blocky

Fine
columnar
flat top

Fine
blocky

Fine
columnar
round top

Fine
blocky

Medium prismlike
20–50 mm

Medium
prismatic

Fine
prismatic

Medium
blocky

Medium
columnar
flat or round top

Fine
prismatic

Medium
blocky

Granular

Fine
<2
mm

Medium
2–5 mm

Coarse
5–10 mm

Platy

Fine <2 mm

Medium 2–5 mm

Large   Coarse >5 mm

Angular blocky

Fine
<10 mm

Medium

Medium
10–20 mm

Coarse
20–50 mm

Very coarse >50 mm

Faces less than five sides

Subangular blocky

Fine
<10 mm

Medium
10–20 mm

Coarse
20–50 mm

Very coarse >50 mm

Faces more than five sides

Figure 43   Types, kinds, and classes of soil structure.

160

Chapter 18

Landform classification

This chapter was prepared by the Landform Mapping Systems Subcommittee of the Canada Soil Survey Committee, 1976: D.F. Acton (chairman), N.F. Alley, R. Baril, A.T. Boydell, J.H. Day, R.J. Fulton, P.K. Heringa, T.M. Lord, J.I. MacDougall, J.L. Nowland, W.W. Pettapiece, E.W. Presant, B. Rochefort, J.A. Shields, R.E. Smith, and M.D. Sudom.

History

A system of landform classification for soil mapping has been desired by soil scientists in Canada for a long time. The Canada Soil Survey Committee (CSSC) adopted a system at a meeting held at the University of Guelph in February 1976. Many aspects of the system came from mapping schemes used by the Geological Survey of Canada for mapping surficial geology. The system also embodies concepts developed initially by R.J. Fulton and later by N.F. Alley while doing terrain mapping in British Columbia. However, the needs of the soil scientist for a terrain or landform classification system are not necessarily compatible with those of the geologist. A national system without any constraints that might be inherent in the system developed for British Columbia was needed. Consequently, the Landform Classification Subcommittee of the CSSC wanted to ensure that the system devised to meet the soil survey requirements in British Columbia, *Terrain Classification System* (Environment and Land Use Committee Secretariat 1976), was acceptable for soil surveys throughout Canada.

Scope

The system of landform classification, developed and presented here, is categorical in nature but does not involve any rigid hierarchy. It is intended to be a field classification system rather than a theoretical taxonomic one.

Landforms in this system are considered to include materials and form. The system emphasizes objectivity whereby the two basic attributes are recognized in terms of their inherent properties rather than on inferred genesis. Genetic terms are frequently used to denote materials and, in some instances, a knowledge of the genesis of a material or form may be required to accurately classify it. The system also attempts to map comprehensively all landforms rather than stress prominent features of importance in interpreting glacial history as is sometimes the case in surficial geological mapping.

The system applies to "local" landforms that are readily represented on maps at scales of 1:50 000 to 1:500 000. These local forms contrast to "regional" landforms that can be represented at scales of 1:1 000 000 or smaller.

In many respects the system is still conceptual in scope and is not parametric because it generally lacks precise limits on the boundaries between classes. It is anticipated that, with further development, parameters can be applied to define classes more rigidly.

The Material category recognizes four groups of materials: unconsolidated mineral, organic, consolidated, and ice. A number of classes of unconsolidated mineral and organic materials have been established, but classes of consolidated materials (bedrock) and ice have not been recognized. Textures of unconsolidated mineral and fiber content of organics are recognized in a category called Material Qualifiers.

Surface Expression, or form, associated with a material or deposit is considered first based on the primary depositional form. Postdepositional forms, essentially erosional, as well as secondary processes are recognized by a category called Modifying Processes.

Finally, a category called Qualifying Descriptors makes possible further qualification of the kinds of materials and the current state of processes as to whether they are active or inactive.

Objectives

1. To provide a standard glossary of local landform terminology for the needs of a soil survey. Other geological terms are defined in the references cited (American Geological Institute 1960*a* and *b*).

2. To provide a system for field collection of landform information in soil surveys involving scales of 1:50 000 and smaller.

3. To provide a system for cataloging local landform information (on-file maps, data banks).

4. To provide a system for recognizing the landform component of the mapping unit used by soil surveys.

Genetic Materials

Materials are classified according to their essential properties within a general framework of their mode of formation. Four groups (components) of materials have been recognized to facilitate further characterization of the texture and the surface expression of the materials. They are unconsolidated mineral, organic, consolidated, and ice components. These groups and the classes established within them are presented below (*see* Figs. 44–49)[1].

Unconsolidated mineral component

The unconsolidated mineral component consists of clastic sediments that may or may not be stratified, but whose particles are not cemented together. They are essentially of glacial or postglacial origin but include poorly consolidated and weathered bedrock. The classes and their definitions follow:

A—Anthropogenic	M—Morainal
C—Colluvial	S—Saprolite
E—Eolian	V—Volcanic
F—Fluvial	W—Marine
L—Lacustrine	U—Undifferentiated

Anthropogenic These materials are artificial or modified by people and include those associated with mineral exploitation and waste disposal.

They include materials constructed or deposited by people or geological materials modified by people so that their physical properties (structure, cohesion, compaction) have been drastically altered. These materials commonly possess a wide range of textures. The process is assumed to be active.

Examples: areas of landfill, spoil heaps, and open-pit mines.

On-site symbols are used for anthropogenic sites where the zone of disturbance is too small to be mapped as an areal unit (*see* Fig. 72).

Colluvial These sediments are massive to moderately well stratified and nonsorted to poorly sorted with any range of particle sizes, from clay to boulders, and blocks that have reached their present position by direct, gravity-induced movement (*see* Fig. 44).

They are restricted to products of mass-wasting whereby the debris is not carried by wind, water, or ice (excepting snow avalanches). Processes include slow displacements such as creep and solifluction and rapid movements such as earth flows, rockslides, avalanches and falls. The process is assumed to be active.

Where colluvial materials are derived from an unconsolidated deposit but overlie a different unit or form a discrete surface expression, they are mapped as colluvial. Colluvial material, which is derived from unconsolidated Quaternary sediments, that overlies and resembles its parent unit is mapped as the parent unit. Colluvial materials exclude those materials deposited at the base of steep slopes by unconcentrated surface runoff or sheet erosion.

Examples: creep, solifluction, earth flows, rockslides, avalanches and falls are processes that produce colluvial material.

Eolian These sediments generally consist of medium to fine sand and coarse silt. They are well sorted, poorly compacted, and may show internal structures such as cross bedding or ripple laminae, or may be massive. Individual grains may be rounded and show signs of frosting (*see* Fig. 45).

---

[1] Some examples of materials and surface expression are shown in Figures 44–71. The photographs were made available through the courtesy of: C.J. Acton (58, 63); D.F. Acton (45, 47, 49, 50, 52, 53, 57, 62); N.F. Alley (44, 46, 64); P. Baurdeau (48); R. Marcoux (56, 67); K. Michalica (66); J.L. Nowland (51); B. Rochefort (65); J.A. Shields (54, 59, 60, 61); C. Tarnocai (68, 69, 70, 71); C. Wang (55).

These materials have been transported and deposited by wind action. In most cases the process is assumed to be inactive.

Examples: dunes, veneers and blankets of sand and coarse silt, and loess but not tuffs.

Fluvial These sediments generally consist of gravel and sand with a minor fraction of silt and rarely of clay. The gravels are typically rounded and contain interstitial sand. Fluvial sediments are commonly moderately to well sorted and display stratification, but massive, nonsorted fluvial gravels do occur. These materials have been transported and deposited by streams and rivers (*see* Fig. 46).

The process is assumed to be inactive.

Examples: channel deposits, overbank deposits, terraces, alluvial fans, and deltas.

Lacustrine These sediments generally consist of either stratified fine sand, silt, and clay deposited on a lake bed; or moderately well sorted and stratified sand and coarser materials that are beach and other nearshore sediments transported and deposited by wave action (*see* Fig. 47).

These materials have either settled from suspension in bodies of standing fresh water or have accumulated at their margins through wave action. The process is assumed to be inactive.

Examples: lake sediments and beaches.

Morainal These sediments generally consist of well-compacted material that is nonstratified and contains a heterogeneous mixture of particle sizes. It commonly comprises a mixture of sand, silt, and clay that has been transported beneath, beside, on, within, and in front of a glacier and not modified by any intermediate agent (*see* Fig. 49).

Examples: basal till (ground moraine), lateral and terminal moraines, rubbly moraines of cirque glaciers, hummocky ice–disintegration moraines, and preexisting, unconsolidated sediments reworked by a glacier so that their original character is largely or completely destroyed.

Saprolite This material is rock and contains a high proportion of residual silts and clays formed by alteration, chiefly by chemical weathering.

The rock remains in a coherent state, interstitial grain relationships are undisturbed, and no downhill movement due to gravity has occurred. The process is assumed to be active.

Examples: rotten rock containing corestones.

Volcanic The deposits consist of unconsolidated pyroclastic sediments. The process is assumed to be inactive.

Examples: volcanic dust, ash, cinders, and pumice.

Marine These unconsolidated deposits of clay, silt, sand, or gravel are well to moderately well sorted and well to moderately stratified (in some places containing shells). They have settled from suspension in salt or brackish water bodies or have accumulated at their margins through shoreline processes such as wave action and longshore drift (*see* Fig. 48).

Nonfossiliferous deposits may be judged marine, if they are located in an area that might reasonably be considered to have contained salt water at the time the deposits were formed. The process is assumed to be inactive.

Undifferentiated This classification is used for a layered sequence of more than three types of genetic material outcropping on a steep erosional escarpment.

This complex class is used where units relating to individual genetic materials cannot be delimited separately at the scale of mapping. It may include colluvium derived from various genetic materials and resting upon a scarp slope.

Organic component

The organic component consists of peat deposits containing >17% organic C (>30% organic matter) by weight. These deposits may be as thin as 10 cm if they overlie bedrock but are otherwise greater than 40 cm and generally greater than 60 cm thick. The classes and their definitions follow:

B—Bog (sphagnum or forest peat)
N—Fen (fen or sedge peat)
O—Organic, undifferentiated
S—Swamp (forest peat)

Bog These deposits consist of sphagnum or forest peat formed in an ombrotrophic environment caused by the slightly elevated nature of the bog. They tend to be disassociated from nutrient-rich ground water or surrounding mineral soils.

Near the surface the materials are usually undecomposed (fibric), yellowish to pale brown in color, loose and spongy in consistence, and entire sphagnum plants are readily identified. These materials are extremely acid (pH <4.5), and have low bulk density (<0.075 mg cm^{-3}) and very high fiber content (>85% unrubbed and ≥40% rubbed). At depths they become darker in color, compacted, and somewhat layered. Bogs are associated with slopes or depressions with a water table at or near the surface in the spring and slightly below it during the remainder of the year. They are usually covered with sphagnum, but sedges may also grow on them. Bogs may be treed or treeless and are frequently characterized by a layer of ericaceous shrubs.

Fen These deposits consist of sedge peat derived primarily from sedges with inclusions of partially decayed stems of shrubs formed in a eutrophic environment due to the close association of the material with mineral-rich waters.

It is usually moderately well to well decomposed, dark brown in color, with fine- to medium-sized fibers but may be well decomposed and black with fine fibers. Decomposition often becomes greater at lower depths. Fen materials are medium acid to weakly alkaline (pH 5.5–7.5), relatively low in fiber (20–80% unrubbed and 2–25% rubbed), and relatively dense (0.075–0.195 mg cm^{-3}). They are associated with relatively open peat lands with a mineral-rich water table that persists seasonally at or very near the surface. The materials are covered with a dominant component of sedges, but grasses and reeds may be associated in local pools. Sphagnum is usually subordinate or absent and the more exacting mosses are common. Often there is much low to medium height shrub cover and sometimes a sparse layer of trees.

Swamp This class is used for forest peat-covered or forest peat-filled areas where the water table is at or above the peat surface. The dominant peat materials are shallow to deep mesic to humic forest and fen peat formed in a eutrophic environment resulting from strong water movement from the margins or other mineral sources.

The peat is usually moderately well to well decomposed and has a dark brown to reddish brown matrix; the more decomposed materials are black in color. It has an amorphous or very fine fibered structure and somewhat layered macrostructure and contains a random distribution of coarse- to medium-sized woody fragments. There may be layers of larger woody particles consisting of stems, roots, and trunks of coniferous tree species. Forest peat materials are usually base saturated and medium acid to weakly alkaline (pH 5.5–7.5); the matrix material is relatively dense (≥0.075 mg cm^{-3}) and density increases with depth. The fiber content is intermediate between sphagnum and fen peats (about 55% unrubbed and 10% rubbed). These materials are associated with stream courses, lake edges, subsurface drainage, glacial depressions, and bog margins. Standing to gently flowing waters occur seasonally or persist for long periods on the surface. The substrate is usually continually waterlogged. The vegetation cover may consist of coniferous or deciduous trees, tall shrubs, herbs, and mosses. In some regions sphagnum mosses may abound.

Consolidated component

The consolidated component consists of tightly packed, indurated materials of bedrock origin. The materials include igneous, metamorphic, sedimentary, and consolidated volcanic rocks. The only class is bedrock (R), which is undifferentiated.

Ice component

The ice component includes areas of snow and ice where evidence of active glacier movement is present within the boundary of the defined unit area. Ice movement is indicated by features such as crevasses, supraglacial moraines, icefalls, and ogives. The only class is ice (I) which is undifferentiated. The process is assumed to be active.

Examples: cirque glaciers, mountain icefields, valley and piedmont glaciers.

Material Modifiers

Material modifiers are used to qualify unconsolidated mineral and organic deposits. Particle-size classes serve to indicate the size, roundness, and sorting of unconsolidated mineral deposits. Fiber classes indicate the degree of decomposition and fiber size of organic materials.

Particle-size classes for unconsolidated materials

The particle-size clases and definitions for unconsolidated materials are as follows:

a—Blocky	l—Loamy
b—Bouldery	p—Pebbly
c—Clayey	r—Rubbly
k—Cobbly	s—Sandy
g—Gravelly	si—Silty

Blocky An accumulation of angular particles greater than 256 mm in size.

Bouldery An accumulation of rounded particles greater than 256 mm in size.

Clayey An accumulation of particles where the fine earth fraction contains 35% or more clay (≤0.002 mm) by weight and particles greater than 2 mm are less than 35% by volume.

Cobbly An accumulation of rounded particles having a diameter of 64–256 mm.

Gravelly An accumulation of rounded particles ranging in size from pebbles to boulders.

Loamy An accumulation of particles of which the fine earth fraction contains less than 35% clay (≤0.002 mm) by weight and less than 70% fine sand and coarser particles. Particles coarser than 2 mm occupy less than 35% by volume.

Pebbly An accumulation of rounded particles having a diameter of 2–64 mm.

Rubbly An accumulation of angular fragments having a diameter of 2–256 mm.

Sandy An accumulation of particles of which the fine earth fraction contains more than 70% by weight of fine sand or coarser particles. Particles greater than 2 mm occupy less than 35% by volume.

Silty An accumulation of particles of which the fine earth fraction contains less than 15% of fine sand or coarser particles and has less than 35% clay. Particles greater than 2 mm occupy less than 35% by volume.

Well-sorted materials are generally described using a single particle-size term; less well sorted and poorly sorted materials are described using two particle-size terms. A subordinate textural component is not generally shown if it constitutes less than 35% of the total volume of the deposit.

Fiber classes for organic materials

The amount of fiber and its durability are important characterizing features of organic deposits in that they reflect on the degree of decomposition of the material. The prevalence of woody materials in peats is also of prime importance. The fiber classes and definitions for organic materials are as follows:

f—Fibric
m—Mesic
h—Humic
w—Woody

Fibric The least decomposed of all organic materials; large amounts of well-preserved fiber(s) are present that are readily identifiable as to their botanical origin. Fibers retain their character upon rubbing.

Mesic Organic material in an intermediate stage of decomposition; intermediate amounts of fiber are present that can be identified as to their botanical origin.

Humic Highly decomposed organic material; small amounts of fiber are present that can be identified as to their botanical origin. Fibers can be easily destroyed by rubbing.

Woody Organic material containing more than 50% woody fibers.

Surface Expression

The surface expression of genetic materials is their form (assemblage of slopes) and pattern of forms. Form as applied to unconsolidated deposits refers specifically to the product of the initial mode of origin of the materials. When applied to consolidated materials, form refers to the product of their modification by geological processes. Surface expression also indicates how unconsolidated genetic materials relate to the underlying unit. Examples of surface expressions of genetic materials are presented in Figures 50-71.

Classes for unconsolidated and consolidated mineral components

The classes and definitions for unconsolidated and consolidated mineral components are as follows:

a—Apron	m—Rolling
b—Blanket	r—Ridged
f—Fan	s—Steep
h—Hummocky	t—Terraced
i—Inclined	u—Undulating
l—Level	v—Veneer

Apron A relatively gentle slope at the foot of a steeper slope and formed by materials from the steeper, upper slope (*see* Figs. 50 and 53).

Examples: two or more coalescing fans, a simple slope.

Blanket A mantle of unconsolidated materials thick enough to mask minor irregularities in the underlying unit but still conforming to the general underlying topography (*see* Fig. 56 and 67).

Examples: lacustrine blanket overlying hummocky moraine.

Fan A fan-shaped form similar to the segment of a cone and having a perceptible gradient from the apex to the toe (*see* Fig. 52).

Examples: alluvial fans, talus cones, some deltas.

Hummocky A very complex sequence of slopes extending from somewhat rounded depressions or kettles of various sizes to irregular to conical knolls or knobs. The surface generally lacks concordance between knolls or depressions. Slopes are generally 9–70% (5–35°) (*see* Figs. 51, 57, and 62).

Examples: hummocky moraine, hummocky glaciofluvial.

Inclined A sloping, unidirectional surface with a generally constant slope not broken by marked irregularities. Slopes are 2–70% (1–35°). The form of inclined slopes is not related to the initial mode of origin of the underlying material.

Examples: terrace scarps, river banks.

Level A flat or very gently sloping, unidirectional surface with a generally constant slope not broken by marked elevations and depressions. Slopes are generally less than 2% (1°) (*see* Fig. 63).

Examples: floodplain, lake plain, some deltas.

Rolling A very regular sequence of moderate slopes extending from rounded, sometimes confined concave depressions to broad, rounded convexities producing a wavelike pattern of moderate relief. Slope length is often 1.6 km or greater and gradients are greater than 5% (3°) (*see* Fig. 59).

Examples: bedrock-controlled ground moraine, some drumlins.

Ridged A long, narrow elevation of the surface, usually sharp crested with steep sides. The ridges may be parallel, subparallel, or intersecting (*see* Figs. 54 and 58).

Examples: eskers, crevasse fillings, washboard moraines, some drumlins.

Steep Erosional slopes, greater than 70% (35°), on both consolidated and unconsolidated materials. The form of a steep erosional slope on unconsolidated materials is not related to the initial mode of origin of the underlying material.

Examples: escarpments, river banks, and lakeshore bluffs.

Terraced A scarp face and the horizontal or gently inclined surface (tread) above it (*see* Fig. 64).

Example: alluvial terrace.

Undulating A regular sequence of gentle slopes that extends from rounded, sometimes confined concavities to broad rounded convexities producing a wavelike pattern of low local relief. Slope length is generally less than 0.8 km and the dominant gradient of slopes is 2–5% (1–3°) (*see* Figs. 55, 60, and 65).

Examples: some drumlins, some ground moraine, lacustrine veneers and blankets over morainal deposits.

Veneer Unconsolidated materials too thin to mask the minor irregularities of the underlying unit surface. A veneer ranges from 10–100 cm in thickness and possesses no form typical of the material's genesis (*see* Figs. 61, 66 and 67).

Examples: shallow lacustrine deposits overlying glacial till, loess cap.

Classes for organic component

The classes and definitions for organic components are as follows:

b—Blanket	h—Horizontal
o—Bowl	p—Plateau
d—Domed	r—Ribbed
f—Floating	s—Sloping

Blanket A mantle of organic materials that is thick enough to mask minor irregularities in the underlying unit but still conforms to the general underlying topography.

Example: blanket bog.

Bowl A bog or fen occupying concave-shaped depressions.

Example: bowl bog.

Figure 44

Figure 45

Figure 46

Figure 47

Figures 44–49   Examples of unconsolidated materials.

Figure 44   Colluvial material.

Figure 45   Eolian material.

Figure 46   Fluvial material.

Figure 47   Lacustrine material.

Figure 48

Figure 49

Figure 50

Figure 51

Figure 48   Thin marine sands over marine clays in the background have been deranged by progressive rotational flow slides in the foreground.

Figure 49   Morainal material.

Figures 50–55   Surface expressions of colluvial, eolian, and fluvial materials.

Figure 50   Colluvial apron at the base of Nahanni Butte, N.W.T.

Figure 51   Hummocky eolian material, active and stabilized sand dunes in Prince Edward Island.

Figure 52

Figure 53

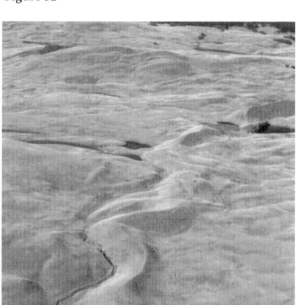

Figure 54

Figure 55

Figure 52   Fluvial fan in the foreground, Carcajou Lake, N.W.T.

Figure 53   Fluvial apron in the midground, Carcajou Canyon, N.W.T.

Figure 54   Ridged and hummocky glaciofluvial material, Kamloops, B.C.

Figure 55   Undulating glaciofluvial material, eastern New Brunswick.

Figure 56

Figure 57

Figure 58

Figure 59

Figures 56–61   Surface expression of morainal materials.

Figure 56   Morainal blanket over undulating bedrock, eastern Quebec.

Figure 57   Hummocky and ridged morainal material in the midground and background, Kamloops, B.C.

Figure 58   Ridged morainal material. The lines of trees mark the swales between parallel ridges, southern Ontario.

Figure 59   Rolling morainal material, southeastern Alberta.

Figure 60

Figure 61

Figure 62

Figure 63

Figure 60   Undulating morainal material, southern Saskatchewan.

Figure 61   Morainal veneer over rolling bedrock, Vancouver Island, B.C.

Figures 62–67   Surface expressions of lacustrine and marine materials.

Figure 62   Hummocky glaciolacustrine material, Biggar, Saskatchewan.

Figure 63   Level lacustrine material, southwestern Ontario.

Figure 64

Figure 65

Figure 66

Figure 67

Figure 64   A lacustrine terrace dissected by streams between a river and hills, Kamloops, B.C.

Figure 65   Undulating marine landform marks the remnants of ancient clay flow slides, Pontiac County, Que.

Figure 66   Thin marine veneer over level bedrock, Grande-Anse, N.B.

Figure 67   Marine veneer and blanket over hummocky bedrock, Montmagny, Que.

172

Figure 68

Figure 69

Figure 70

Figure 71

Figures 68–71   Surface expressions of bogs and fens.

Figure 68   The domed bog in the midground has mainly sphagnum vegetation, Sibbeston Lake, N.W.T.

Figure 69   Plateaus in this bog are marked by the light brown vegetation with sparse tree cover. The flat, reddish brown areas are sedge-covered fens, Norman Wells, N.W.T.

Figure 70   The horizontal fen in the foreground is dominated by sedge vegetation, Manitoba.

Figure 71   The ribbed fen has sedge vegetation broken by low ridges where spruce trees grow, Fort Simpson, N.W.T.

173

Domed A bog with an elevated, convex, central area much higher than the margin. Domes may be abrupt (with or without a frozen core) or gently sloping or have a stepped surface (*see* Fig. 68).

Examples: palsa, peat mound, raised bog.

Floating A level organic surface associated with a pond or lake and not anchored to the lake bottom.

Example: floating fen.

Horizontal A flat peat surface not broken by marked elevations and depressions (*see* Fig. 70).

Examples: flat bog, horizontal fen.

Plateau A bog with an elevated, flat, central area only slightly higher than the margin (*see* Fig. 69).

Examples: peat plateau, bog plateau, polygonal peat plateau.

Ribbed A pattern of parallel or reticulate low ridges associated with fens (*see* Fig. 71).

Examples: string fen, ribbed fen, net fen.

Sloping A peat surface with a generally constant slope not broken by marked irregularities.

Example: sloping fen.

Slope

A set of slope classes has been provided to make it possible to quantify the dominant but not necessarily most abundant slopes within a mapped unit of a local landform. There are 10 slope classes. Each is defined in terms of percent and degrees. Measurements are to the nearest tenth in the two lowest classes.

Slope class	Percent slope	Approximate degrees	Terminology
1	0–0.5	0	level
2	>0.5–2	0.3–1.1	nearly level
3	>2–5	>1.1–3	very gentle slopes
4	>5–10	>3–5	gentle slopes
5	>10–15	>5–8.5	moderate slopes
6	>15–30	>8.5–16.5	strong slopes
7	>30–45	>16.5–24	very strong slopes
8	>45–70	>24–35	extreme slopes
9	>70–100	>35–45	steep slopes
10	>100	>45	very steep slopes

Modifying Processes

Terms that describe the geological processes that have modified or are currently modifying genetic materials and their surface expression are considered within the modifying processes category of the system.

These modifiers are to be used where a relatively large part of the map unit is modified. On-site symbols can be used to indicate that only a relatively small part of a map unit is modified.

The assumed common process status (active, inactive) is specified in the definition of each modifier. Where this status varies from the assumed state, it must be qualified in the description.

A—Avalanched	K—Karst modified
B—Beveled	N—Nivated
C—Cryoturbated	P—Piping
D—Deflated	S—Soliflucted
E—Eroded (channeled)	V—Gullied
F—Failing	W—Washed
H—Kettled	

Avalanched Describes slopes modified by frequent avalanche activity.

An avalanche is a large mass of snow, ice, soil, or rock, or mixtures of these materials, falling or sliding rapidly under the force of gravity. The process is assumed to be active.

Examples: avalanche cones, avalanche tracks or chutes.

Beveled Describes a surface cut or planed by running water but not underlain by fluvial materials.

Beveled applies to river-cut terraces in bedrock and river terraces cut into till or lacustrine silts. The process is assumed to be inactive.

Example: river-cut terrace in bedrock.

Cryoturbated Describes a surface modified by processes of frost action.

It includes surfaces produced by the stirring, churning, modification and other disturbances of soil that result from frost action. This involves frost heaving and differential and mass movements, which produce patterned ground. The process is assumed to be active. Processes involving downslope movements of material overlying a frozen layer are excluded from this modifier and considered more specifically as soliflucted.

Examples: sorted nets, stripes, unsorted circles, earth hummocks.

Deflated Describes a surface modified by the sorting out, lifting, and removal of loose, dry, fine-grained particles (clay and silt sizes) by the turbulent, eddy action of the wind. The process is assumed to be inactive.

Example: deflated lacustrine terrace.

Eroded (channeled) Describes a surface modified by a series of abandoned channels.

The term applies to fluvial plains, terraces, and fans. The process is assumed to be inactive.

Example: abandoned channels on alluvial terrace.

Failing Describes a surface modified by the formation of tension fractures or by large consolidated or unconsolidated masses moving slowly downslope.

Colluvial processes resulting in shallow surface movements are not described as failing. The process is only active.

Example: slumps.

Kettled Describes a surface, deposit, or feature modified by depressions left by melting ice blocks.

Depressions can be formed by the melting blocks of ice buried in glaciofluvial, glaciolacustrine, or glacial till materials. Kettle depressions usually have steep sides and are bound by an abrupt convex break of slope. They occur in a variety of shapes and sizes from round basins to branching valleys. The process is assumed to be inactive.

Examples: pitted outwash and lacustrine, knob and kettle topography.

Karst modified Describes a modification of carbonate and other rocks whose surfaces are marked by features of collapse and solution and also where the rocks are overlain by unconsolidated materials that show posthumous collapse depressions. The process is assumed to be active.

Examples: sinkholes, dolines, uvalas.

Nivated Describes a surface modified by frost action, erosion, and mass wasting beneath and around a snowbank so as to produce transverse, longitudinal, and circular hollows.

Examples: nivation terraces in colluvium, nivation hollow.

Piping Describes a surface modified by small hollows, commonly aligned along routes of subsurface drainage, and resulting from the subsurface removal of particulate matter in unconsolidated materials.

It occurs most commonly in lake silts but may also affect alluvium, loess, and volcanic ash. The process is assumed to be active.

Example: piping in silty lacustrine terrace.

Soliflucted Describes a surface modified by the process of slow gravitational movement downslope of saturated, nonfrozen earth material behaving apparently as a viscous mass over a surface of frozen ground.

Soliflucted surfaces are commonly associated with processes of cryoturbation and nivation occurring in alpine and subalpine areas. The process is assumed to be active.

Examples: lobes, stripes, sheets, terracettes.

Gullied Describes a surface modified by fluvial erosion resulting in the development of parallel and subparallel, steep-sided, and narrow ravines in both consolidated and unconsolidated materials. The process is assumed to be active.

Example: gullied lacustrine terrace.

Washed Describes a surface, deposit, or feature modified by wave action in a body of standing water resulting in lag deposits, beaches of lag materials, and wave-cut platforms.

Washed surfaces occur most commonly in areas of former marine inundation or glacial lakes. Active washing occurs along present shorelines. The process is assumed to be active.

Example: terrace or beach that is cut or deposited on a morainal blanket.

Qualifying Descriptors

A number of descriptors have been introduced to qualify either the genetic materials or the modifying process terms. The descriptors add information about the mode of formation or depositional environment. They also qualify the status of the genetic and modifying processes. Included in the definitions of the categories are statements concerning the commonly assumed status of their processes. Where the process status is contrary to the common assumption, it is indicated.

Clastic   G Glacial, Glaciofluvial, etc.

Process   A Active, I Inactive

Glacial   Used to qualify nonglacial genetic materials or process modifiers where there is direct evidence that glacier ice exerted a strong but secondary or indirect control upon the mode of origin of the materials or mode of operation of the process. The use of this qualifying descriptor implies that glacier ice was close to the site of the deposition of a material or the site of operation of a process.

Glaciofluvial   Used only where fluvial materials show clear evidence of having been deposited either directly in front of, or in contact with, glacier ice. At least one of the following characteristics must be present:

- kettles, or an otherwise irregular (possibly hummocky or ridged) surface that resulted from the melting of buried or partially buried ice, e.g., pitted outwash, knob and kettle topography
- either slump structures or their equivalent topographic expression, or both, indicating partial collapse of a depositional landform due to melting of supporting ice, e.g., kame terrace, delta kame
- ice-contact and molded forms such as gravelly or sandy crevasse fillings and eskers
- nonsorted and nonbedded gravel of an extreme range of particle sizes, such as results from very rapid aggradation at an ice front, e.g., ice-contact gravels
- flowtills.

Glaciolacustrine   Used where there is evidence that the lacustrine materials were deposited in contact with glacial ice. One of the following characteristics must be present:

- kettles or an otherwise irregular surface that is neither simply the result of normal settling and compaction in silt nor the result of piping
- slump structures resulting from loss of support caused by melting of retaining ice
- presence of numerous ice-rafted stones in the lacustrine silts.

Glaciomarine   Used only where it is clear that materials of glacial origin were laid down in a marine environment, i.e., deposits settled through the waters from melting, floating ice and ice shelves. Sediments may be poorly sorted and poorly stratified to nonsorted and massive; shells present will generally be whole and in growth positions.

Meltwater channels   Used to indicate the presence of glacial meltwater channels in a unit where they are either too small or too numerous, or both, to show individually by an on-site symbol.

Active   Used to indicate any evidence of the recurrent nature of a modifying process or of the contemporary nature of the process forming a genetic material.

Inactive   Used to indicate that there is no evidence of the modifying process recurring and that the processes that formed the genetic materials have ceased.

Mapping Conventions

Map symbols

The following example illustrates a system for ordering symbols used in map edits. It assumes that all components of the system (genetic materials and their particle-size or fiber class, surface expression and related slope, modifying processes, and qualifying descriptors) are to be used.

$$\frac{tG^Qe^Q / tG^Qe^Q 1 - P^Q}{G^Qe}$$

$t\,G^Q\,e^Q$ is the dominant ($\geq 50\%$) surficial material and $t\,G^Q\,e^Q\,1 - P^Q$ is the subdominant material. Rarely will materials occupying less than 15% of the map area be recognized in the edit. (On-site symbols [Fig. 72] provide a mechanism for depicting many of these).

t—genetic material modifier (particle size of clastic materials and fiber content of organic materials).
G—genetic materials
e—surface expression
l—slope qualifiers (numeric)
P—modifying processes
Q—qualifying descriptor (superscript)
G^Qe—in the denominator represents an underlying stratigraphic unit.

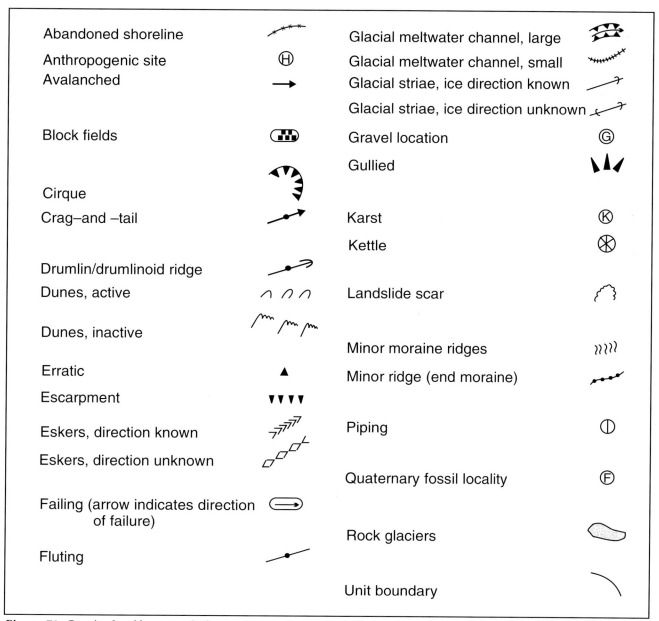

Figure 71 On-site landform symbols.

Composite units

Not all terrain can be presented as simple units because terrain units commonly occur that are of small areal extent and cannot be delimited individually at the scale of mapping. Consequently, a system of composite units is employed whereby up to three types of terrain may be designated within a common unit boundary. The relative amounts of each terrain type are indicated by the use of the symbols =, /, and //. The components are always indicated in decreasing order of abundance.

= components on either side of this symbol are about equal; each represents 45–55% of the area.

/ the component in front of this symbol is more abundant than the one that follows; the first represents 55–70% of the area and the second 30–45%.

// the component in front of this symbol is considerably more abundant than the one that follows; the first represents 70–90% of the area and the second 10–30%.

Stratigraphic data

Stratigraphic data may be presented to supplement the surficial data. Stratigraphic information should be given for veneers and blankets. Also, where the depth of the surface unit is such that the nature of the underlying unit may be important or where it is deemed necessary to show the character of the underlying unit, or both, then stratigraphic information should be shown.

For map presentation a horizontal bar is used to separate the components that are arranged in stratigraphic order. Surface expressions may be attached to underlying units if appropriate.

If the type of unconsolidated material underlying a blanket of different genetic material is not known, then only the surface expression of that underlying material is given. For example, Mb/h indicates hummocky unconsolidated material of unknown specific type underlying a blanket of morainal materials.

On-site symbols

On-site symbols or map symbols (*see* Fig. 72) are used to describe features or processes in the terrain that express either a limited (by scale), areal function or are simply point observations. These may be linear features such as eskers or moraine ridges, site-specific information such as gravel locations or kettle holes, or to add details of Quaternary history such as striae, glacial meltwater channels, or abandoned shorelines.

The size of the on-site symbols varies with the type of symbol. For example, the symbols that connote an areal extent such as failing or block fields vary in size, whereas those that are point observations and have no relation to areal extent, such as fossil locality or karst, will be of a standard size. The symbols that have linear connotations, such as eskers, gullying, or end moraines, vary in length but are of standard width.

References

Agriculture Canada. 1974. A system of soil classification for Canada. Agric. Can. Publ. 1455. Revised. Information Canada, Ottawa, Ont. 255 pp.

Agriculture Canada Expert Committee on Soil Survey. 1983. The Canada soil information system (CanSIS) manual for describing soils in the field. 1982 Revised. Agriculture Canada, Ottawa, Ont. 164 pp.

Agriculture Canada Expert Committee on Soil Survey. 1987. The Canadian system of soil classification. 2nd ed. Agric. Can. Publ. 1646. 164 pp.

American Geological Institute. 1960a. Glossary of geology and related sciences. 2nd ed. American Geological Institute, Washington, D.C. 397 pp.

American Geological Institute. 1960b. Dictionary of geological terms. Doubleday, Garden City, N.Y. 545 pp.

Baldwin, M.; Kellogg, C.E.; Thorp, J. 1938. Soil classification. Pages 979–1001 in United States Department of Agriculture. Soils and men: Yearbook of agriculture 1938. U.S. Government Printing Office, Washington, D.C.

Boelter, D.H. 1969. Physical properties of peats as related to degree of decomposition. Soil Sci. Soc. Am. Proc. 33:606–609.

Brierley, J.A.; Mermut, A.R.; Stonehouse, H.B. 1996. A new order in the Canadian system of soil classification: Vertisolic soils. Saskatchewan Land Resource Unit, Saskatoon, Sask. CLBRR Cont. No. 96-11. 76 pp.

Brown, R.J.E.; Kupsch, W.O. 1974. Permafrost terminology, NRCC Tech. Memo. 111. 62 pp.

Canada Department of Agriculture. 1976. Glossary of terms in soil science. Agric. Can. Publ. 1459. Revised. 44 pp.

Canada Soil Survey Committee. 1970. Proceedings of the eighth meeting of the Canada Soil Survey Committee of Canada. Mimeographed report of the Canada Department of Agriculture, Ottawa, Ont.

Canada Soil Survey Committee. 1973. Proceedings of the ninth meeting of the Canada Soil Survey Committee of Canada. Mimeographed report of the Canada Department of Agriculture, Ottawa, Ont. 354. pp.

Canada Soil Survey Committee. 1974. Proceedings of the tenth meeting of the Canada Soil Survey Committee of Canada. Mimeographed report of the Canada Department of Agriculture, Ottawa, Ont. 107 pp.

Canada Soil Survey Committee. 1976. Proceedings of the eleventh meeting of the Canada Soil Survey Committee of Canada. Mimeographed report of the Canada Department of Agriculture, Ottawa, Ont.

Canada Soil Survey Committee, Subcommittee on Soil Classification. 1978. The Canadian system of soil classification. Can. Dep. Agric. Publ. 1646. Supply and Services Canada, Ottawa, Ont. 164 pp.

Carter, M.R., ed. 1993. Soil sampling and methods of analysis. Canadian Society of Soil Science. 823 pp.

Clayton, J.S.; Ehrlich, W.A.; Cann, D.B.; Day, J.H.; Marshall, I.B. 1977. Soils of Canada. Agric. Can. Publ. 1544. 2 vols.

Cline, M.G. 1949. Basic principles of soil classification. Soil Sci. 67:81–91.

Cline, M.G. 1961. The changing model of soil. Soil Sci. Soc. Am. Proc. 25:442–446.

Cline, M.G. 1963. Logic of the new system of soil classification. Soil Sci. 96:17–22.

De Bakker, H. 1970. Purposes of soil classification. Geoderma 4:195–208.

Deckers, J.; Spaargaren, O.C.; Nachtergaele, F.O. eds. 1998. World Reference Base for Soil Resources. International Society of Soil Science Working Group RB, 1st ed., Vols 1 and 2, KULeuven Academic Press, Acco. 93 and 96 pp.

Ellis, J.H. 1932. A field classification of soils for use in soil survey. Sci. Agric. 12:338–345.

Ellis, J.H. 1971. The birth of agronomic research at the Manitoba Agricultural College. 181 pp.

Environment and Land Use Committee Secretariat. 1976. Terrain classification system. Government of British Columbia, Victoria, B.C. 54 pp.

Food and Agriculture Organization of the United Nations. 1985. FAO-UNESCO soil map of the world. Revised legend. UNESCO, Rome. 115 pp.

Fox, C.A. 1985. Characteristics of the Folisolic soils of British Columbia. Pages 205–232 in J.A. Shields and D.J. Kroetsch, eds. Proceedings, Sixth Annual Meeting, Expert Committee on Soil Survey, Guelph, Ont., 26–30 November 1984. Research Branch, Agriculture Canada, Ottawa, Ont.

Fox, C.A.; Trowbridge, R.; Tarnocai, C. 1987. Classification, macromorphology and chemical characteristics of Folisols from British Columbia. Can. J. Soil Sci. 67:765–778.

Glinka, K.D. 1927. The great soil groups of the world and their development. (Translated from the German by C.F. Marbut.) Edwards Brothers, Ann Arbor, Mich. 235 pp.

Harris, S.A., French, H.M.; Heginbottom, J.A.; Johnston, G.H.; Ladanyi, B.; Sego, D.C.; van Everdingen, R.O. 1988. Glossary of permafrost and related ground-ice terms. National Research Council of Canada, Associate Committee on Geotechnical Research, Ottawa, Technical Memorandum No. 142. 156 pp.

Jenny, H. 1961. E.W. Hilgard and the birth of modern soil science. Collana Della Rivista Agrochimica, Pisa, Italy. 144 pp.

Joel, A.H. 1926. Changing viewpoints and methods in soil classification. Sci. Agric. 6:225–232.

Kellogg, C.E. 1941. The soils that support us. Macmillan, New York, N.Y. 309 pp.

Knox, E.G. 1965. Soil individuals and soil classification. Soil Sci. Soc. Am. Proc. 29:79–84.

McKeague, J.A., ed. 1978. Manual on soil sampling and methods of analysis. Second edition, Soil Research Institute, Agriculture Canada, Ottawa, Ont. 212 pp.

McKeague, J.A.; Wang, C.; Tarnocai, C.; Shields, J.A. 1986. Concepts and classification of Gleysolic soils in Canada. Research Branch, Agriculture Canada, Ottawa, Ont. 38 pp.

National Soil Survey Committee. 1945. Proceedings of the first conference of the National Soil Survey Committee of Canada. Mimeographed report of the Canada Department of Agriculture, Ottawa, Ont. 148 pp.

National Soil Survey Committee. 1948. Proceedings of the second meeting of the National Soil Survey Committee of Canada. Mimeographed report of the Canada Department of Agriculture, Ottawa, Ont.

National Soil Survey Committee. 1955. Proceedings of the third conference of the National Soil Survey Committee of Canada. Mimeographed report of the Canada Department of Agriculture, Ottawa, Ont. 118 pp.

National Soil Survey Committee. 1960. Proceedings of the fourth meeting of the National Soil Survey Committee of Canada. Mimeographed report of the Canada Department of Agriculture, Ottawa, Ont. 42 pp.

National Soil Survey Committee. 1963. Proceedings of the fifth meeting of the National Soil Survey Committee of Canada. Mimeographed report of the Canada Department of Agriculture, Ottawa, Ont. 92 pp.

National Soil Survey Committee. 1965. Proceedings of the sixth meeting of the National Soil Survey Committee of Canada. Mimeographed report of the Canada Department of Agriculture, Ottawa, Ont.

National Soil Survey Committee. 1968. Proceedings of the seventh meeting of the National Soil Survey Committee of Canada. Mimeographed report of the Canada Department of Agriculture, Ottawa, Ont. 216 pp.

Pihlainen, J.A.; Johnston, G.H. 1963. Guide to a field description of permafrost for engineering purposes. NRCC Tech. Memo. 79. 21 pp.

Ruhnke, G.N. 1926. The soil survey of southern Ontario. Sci. Agric. 7:117–124.

Simonson, R.W. 1968. Concept of a soil. Adv. Agron. 20:1–47.

Soil Survey Staff. 1951. Soil survey manual. U.S. Dep. Agric. Handbk. No. 18. U.S. Government Printing Office, Washington, D.C. 503 pp.

Soil Survey Staff. 1960. Soil classification, a comprehensive system, 7th approximation. U.S. Government Printing Office, Washington, D.C. 265 pp.

Soil Survey Staff. 1975. Soil taxonomy. U.S. Dep. Agric. Handbk. No. 436. U.S. Government Printing Office, Washington, D.C. 754 pp.

Soil Survey Staff. 1992. Keys to Soil Taxonomy, 5th editon. SMSS Technical Monograph No. 19. Blacksburg, Virginia: Pocahontas Press, Inc. 556 pp.

Tarnocai, C. 1985. Soil classification. Pages 283–294 in J.A. Shields and D.J. Kroetsch, eds. Proceedings, sixth annual meeting, Expert Committee on Soil Survey, Guelph, Ont., 26–30 November 1984. Research Branch, Agriculture Canada, Ottawa, Ont.

Trowbridge, R. 1981. Second progress report of the working group on Organic horizons, Folisols, and humus form classification. British Columbia Ministry of Forests, Victoria, B.C. 84 pp.

Trowbridge, R.; Luttmerding, H.; Tarnocai, C. 1985. Report on Folisolic soil classification in Canada. Pages 180–204 in J.A. Shields and D.J. Kroetsch, eds. Proceedings, sixth annual meeting, Expert Committee on Soil Survey, Guelph, Ont., 26–30 November 1984. Research Branch, Agriculture Canada, Ottawa, Ont.

Working Group on Soil Survey Data. 1975. The Canadian Soil Information System (CanSIS) manual for describing soils in the field. Soil Res. Inst., Can. Dep. Agric. Ottawa, Ont. 170 pp.

Index

Subgroups are listed in alphabetical order under the great group name.